厨房情景英语

第二版

华路宏 钟 文 主编

图书在版编目(CIP)数据

厨房情景英语 / 华路宏，钟文主编. —2 版. —杭州：浙江工商大学出版社，2018.2(2023.8 重印)
ISBN 978-7-5178-2594-4

Ⅰ. ①厨… Ⅱ. ①华… ②钟… Ⅲ. ①厨房—英语—高等职业教育—教材 Ⅳ. ①TS972.11

中国版本图书馆 CIP 数据核字(2018)第 009062 号

厨房情景英语(第二版)

华路宏　钟　文 主编

责任编辑　王　英
封面设计　林朦朦
责任印制　包建辉
出版发行　浙江工商大学出版社
(杭州市教工路 198 号　邮政编码 310012)
(E-mail:zjgsupress@163. com)
(网址:http://www. zjgsupress. com)
电话:0571-88904980,88831806(传真)
排　　版　杭州朝曦图文设计有限公司
印　　刷　广东虎彩云印刷有限公司绍兴分公司
开　　本　710mm×1000mm　1/16
印　　张　14.75
字　　数　240 千
版 印 次　2018 年 2 月第 1 版　2023 年 8 月第 5 次印刷
书　　号　ISBN 978-7-5178-2594-4
定　　价　35.00 元

浙江工商大学出版社营销部邮购电话　0571-88904970

序

国际化是我国改革开放后旅游业发展的显著特征。最近十来年，伴随着经济的增长，以及和西方国家饮食交流的日益密切，在各大城市甚至包括一些三四线城市，西餐消费渐渐走近寻常百姓。这直接导致了高素质高技能西餐人才的短缺，由此也直接带动了我国西餐职业教育的蓬勃发展，高职院校西餐工艺及相关专业的招生数量逐年递增，西餐工艺毕业生就业出现了供不应求的局面。

教材建设是专业建设的基础性工作，是人才培养的必备条件。目前，国内开设西餐教学的高职院校已逾50所，而专门针对高职层次西餐教学的教材并不完善。浙江旅游职业学院作为国内较早开设西餐工艺专业的高职院校，拥有一批知名的西餐教师，他们不但有理论，而且拥有丰富的行业经验。浙江旅游职业学院的西餐工艺专业系全国唯一一个通过世界旅游组织旅游教育质量认证的烹饪类专业，2010年被纳入国家示范性骨干高职院校重点建设专业，2011年获批中央财政支持“提升专业服务产业发展能力”建设项目。在浙江旅游职业学院国家骨干院校建设期间(2010—2012)，西餐工艺专业实行了系列教学改革，并取得了不俗的成绩：构建并持续推行了“师资联动、文化联动、基地联动、产学联动”的“四联动”育人模式；以国际化视野培养人才，在迪拜、阿布扎比、中国澳门等地及意大利哥诗达邮轮上实习或就业的学生占专业总人数的20%以上。所以，西餐工艺专业教师承担系列教材的编撰任务，既是建设国家骨干项目的要求，也是骨干院校建设人才培养经验共享的体现。

本次出版的《西餐工艺实训教程》《西点工艺》《厨房情景英语》和《西餐烹饪原料》四本教材，是系列中的一部分，主要用于西餐工艺专业核心课程的教学。教材编写根据教育部颁布的《关于全面提高高等职业教育教学质量的若干意见》(〔2006〕16号文件)精神，遵循“以就业为导向、工学结合”的人才培养指导思想。

综观本系列教材，我认为它有六个方面的特点。(1)“实用、够用”，符合高职高专教育实际。根据高职高专教育重理论更重实践的特点，坚持“实用、够用”原则，结合高职高专学生的知识层次，准确把握教材的内容体系。(2)校企合作，体现工学结合的思路。教材编写过程中与企业进行多方面的合作，教材体例突出项目化和任务型，教材内容与岗位需求做到无缝对接。(3)点面结合，信息量大又重点突出。教材在内容的取舍上，力求精选，不强调面面俱到，注重实用性与典型性的结合，力求保证学生在有限的课时内掌握必备知识，内容丰富，重点突出。教材为学生提供了对应的网络、书刊等资讯，便于学生课余查找和学习，有利于学生拓宽知识面。(4)图文并茂，便于学习认知。所有教材都注重图文并茂，便于学生较直观地认知，有助于学生较快把握各知识点，能够加深记忆，增强学习效果。(5)强化英语，紧扣西餐专业特点。关键知识点都采用中英文对照形式，使学生全方位地掌握专业英语，满足西餐从业人员的英语能力要求。(6)适应面广，满足多专业教学需求。本套教材注重西餐理论知识的普及，突出实践应用，可以满足西餐工艺、酒店管理、餐饮管理与服务、厨政管理等多个专业的教学需求。

浙江旅游职业学院副院长、教授　徐云松

2013 年 6 月

Contents

Chapter 1

Kitchen Introduction and Verbs

Unit 1 Titles Used in the Kitchen

Learning goals

To know the titles used in the kitchen.

To describe the job responsibilities in the kitchen.

Vocabulary

cook [kʊk] *n.* 厨师
chef [ʃef] *n.* 厨师；主厨
executive [ɪgˈzekjətɪv] *adj.* & *n.* 执行的；决策者
pastry [ˈpeɪstri] *n.* 面点
assistant [əˈsɪstənt] *adj.* & *n.* 助理的；助理
vegetable [ˈvedʒtəbl] *n.* 蔬菜；植物
butcher [ˈbʊtʃə(r)] *n.* & *vt.* 屠夫；屠宰
sauce [sɔːs] *n.* & *vt.* 酱汁；给……调味
soup [suːp] *n.* 汤；羹
grill [grɪl] *n.* & *vt.* 烤架；在(烤架上)烤
roast [rəʊst] *n.* & *vt.* 烤肉；烘烤

staff [stɑːf] *n.* 全体人员
aboyeur [ɑːbwɑːˈjuː] *n.* (法语)跑堂喊菜的人
commis [ˈkɒmi] *n.* (法语)小职员;副手
apprentice [əˈprentɪs] *n.* 学徒;见习生
clerk [klɑːk] *n.* 职员;办事员
relief [rɪˈliːf] *n.* 替代
pantryman [ˈpæntrɪmən] *n.* 配餐员;司膳总管;司膳总管助理
potman [ˈpɒtmən] *n.* 酒馆的侍者
porter [ˈpɔːtə(r)] *n.* 搬运工;(列车)服务员;杂务工
steward [ˈstjuːəd] *n.* 乘务员;管家;干事;管理员

Titles used in the kitchen

executive chef	行政总厨师长
assistant chef	行政总厨师长助理
sous-chef	副厨师长，指具体负责并干活的厨师长
larder chef	负责烹饪各种肉类的厨师长
pastry chef	负责烹饪各种面点的厨师长
vegetable chef	负责烹饪各种蔬菜的厨师长
butcher	屠夫,负责屠宰各种禽类的人
grill cook	负责在烤架上烤炙肉类的厨师
sauce cook	负责调制酱汁的厨师
soup cook	负责烹饪各种汤的厨师
fish cook	负责烹饪鱼类的厨师
roast cook	负责在烤箱内烧烤肉类的厨师
breakfast cook	负责烹饪早餐的厨师
night cook	晚上上班的厨师
staff cook	负责烹饪员工伙食的厨师
relief cook	替班厨师,候补厨师。此人是个多面手,任何部门的厨师因休假或病假不能上班,替班厨师就可以补上
commis	助理厨师,是主厨的副手
aboyeur	跑堂喊菜的服务员,他忙于餐厅和厨房之间,把客人的点菜单送给厨师们

kitchen clerk　厨房会计，负责厨房的一切文书工作
pantryman　负责管理配膳室（食品小贮藏室），不是厨师，不烹饪
potman　负责擦洗大深锅的人

Dialogue

Directions: Practice the following dialogue with your partner.
(*S*=*Susan*, *C*=*commis*.)

S: Excuse me, where is the executive chef's office?

C: Go straight and turn right at the end of the hallway. It's on your left-hand side.

S: Thank you. I just want to ask him about the salary.

C: OK. But maybe he is at the beverage cooler now.

S: How long will he be back or shall I go to the beverage cooler to find him?

C: He will come back in 20 minutes. You can wait in his office. I can guide you.

S: Thank you so much!

C: You are welcome.

Activity 1

Task 1　Try to write the names in English below the pictures.

1. ______________

2. ______________

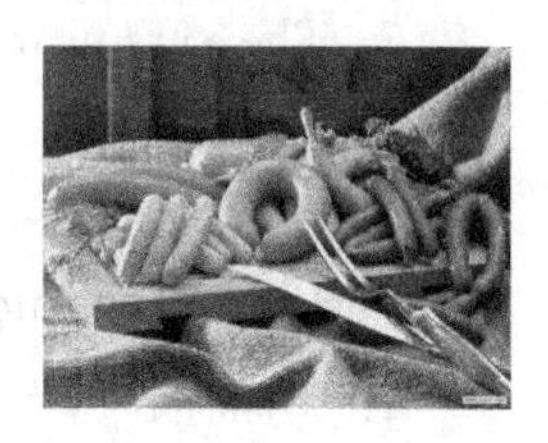

3. ______________

4. ______________ 5. ______________ 6. ______________

7. ______________ 8. ______________

Task 2 Guessing—who are they?

1. The person who can relieve everyone.

2. The person who works between the cook and guests, and carries orders to the cook.

3. The person who makes the breakfast.

4. The person who washes pans and pots.

5. The person who puts the meat into the oven.

6. The person who is responsible for the pantry.

7. The person who cuts the meat and slaughters the animals.

Activity 2

Task 1 Translate the following sentences into English.

1. 请问你在厨房里是做什么的?

____

2. 请问学徒一天在厨房里要工作多久?

____

3. 蔬菜厨师主要负责烹饪各类蔬菜菜肴。

____

4. 你是如何找到洗碗工这份工作的?

____

5. 布莱恩是一名屠夫,他每天要在厨房里屠宰各种家禽。

____

Task 2　Role play.

Setting:招聘行政总厨

Work in groups, role play in the setting of recruiting an executive chef. The scene is set as below:

- up to 22,000 pounds per year
- 28 days holidays + benefits
- bridge pub, west London

Activity 3

Task　Reading Comprehension.

Directions: Work in pairs, read the passage and discuss with your partner about the pastry chef's responsibility.

Pastry Chef

A pastry chef is a station chef in a professional kitchen, skilled in the making of pastries, desserts, breads and other baked goods. They are employed in large hotels, bistros, restaurants, bakeries, and some cafés. Pastry chefs use a combination of culinary ability and creativity in baking, direction, and flavoring with ingredients. Many baked goods require a lot of time and focus. Presentation is an important part of pastry and dessert preparation. The job is often a physically demanding job that requires lots of work with your hands and long hours on your feet and can be stressful with hours that start in the early morning. They are also responsible for creating new recipes to put on the menu. Pastry baking is usually held in a slightly separate part from the main kitchen. This section of the kitchen is in charge of making pastries, desserts, and other baked goods.

Unit 2　Kitchen Safety and Kitchen Hygiene

Learning goals

To know some rules of kitchen safety and kitchen hygiene.

To know some working procedures.

To learn to ask and tell what should be done.

To know dangerous things in the kitchen.

To know the kitchen floor.

Vocabulary

hygiene　[ˈhaɪdʒiːn] *n.* 卫生;卫生学

washer section 锅具清洗区

beverage cooler 饮料冷库

kitchen store 厨房贮藏室

pick up area 备菜间

scullery section 餐具洗涤区

butchery　[ˈbʊtʃəri] *n.* 屠杀;屠场;屠宰业

rodent　[ˈrəʊdnt] *n.* 啮齿动物

insect　[ˈɪnsekt] *n.* 昆虫;虫子

rat　[ræt] *n.* & *v.* 变节者;老鼠;捕鼠

bacteria　[bækˈtɪərɪə] *n.* (复数)细菌

cockroach　[ˈkɒkrəʊtʃ] *n.* 蟑螂

poison　[ˈpɔɪzn] *n.* & *vt.* & *adj.* 毒药;毒害;有毒的;危险的

sanitize　[ˈsænɪtaɪz] *vt.* 采取卫生措施使其安全;消毒

antiseptic　[ˌæntiˈseptɪk] *n.* & *adj.* 杀菌剂;防腐剂; 杀菌的;防腐的

detergent　[dɪˈtɜːdʒənt] *n.* & *adj.* 清洁剂;用于清洗的

extinguisher　[ɪkˈstɪŋgwɪʃə] *n.* 灭火器

Dialogue 1 Kitchen hygiene

(*A=Apprentice*, *C=Chef.*)

A: I am going to work in this kitchen. What rules shall I know then?

C: You should pay attention to the kitchen hygiene and kitchen safety.

A: Could you tell me some details about kitchen hygiene and kitchen safety?

C: Certainly, there are many rats and cockroaches in the kitchen for there is plenty of food in the kitchen.

A: What shall we do?

C: Don't panic. Just poison them or trap them and everything in the kitchen should be sanitized. And also the bacteria may get in food.

A: How do microbes or bacteria get in food?

C: Usually microbes get into the food from the hands of people working in the kitchen, mostly from our hands. So we should always wash our hands.

A: Got it. What kinds of foods are most dangerous?

C: Protein foods, like meat, poultry, eggs, fish, diary products, etc.

A: How can we stop food poisoning?

C: Never leave food outside the refrigerator for more than two hours and keep refrigerated food below 40℉.

Dialogue 2 Kitchen safety

A: Chef, what is the most dangerous thing in the kitchen?

C: Safety is the number one priority in the kitchen. Let me show you some dangerous things and the ways in which we could avoid danger in the kitchen.

A: Yes, I will listen carefully.

C: Firstly, we should pay attention to the "fire". Oily rags are dangerous. Wash them in soapy water or throw them away. Using deep fryers is dangerous, so watch out for smoke, foam or bubbles. You should learn how to put out an oil or grease fire. You need to have carefully supervised practice in putting out oil and grease fires with a proper type of chemical fire extinguisher. Do not use water on an electrical fire!

A: I think I have to be taught the knowledge about it.

C: Secondly, you should pay attention to "burning hurt". Hot oil and melted fat can easily burn you. If you are not careful, cooking with grease and oil is quite dangerous. They can quickly start a big kitchen fire. By the way, damaged electrical wires are also very dangerous. Do not use any damaged electrical equipment.

A: Oh, really? I am afraid of being burned, I'll be careful.

C: Last but not least, you need avoid slipping or tripping. Falling down can be quite dangerous. If you are carrying something hot, you may get badly burned; if you are carrying something made out of glass or something sharp, you may get a deep cut because of slippery.

A: I think I'd better be more careful. Thank you so much!

Activity 1

Task Fill in the blanks in the sentences with the verbs given if you want to keep the kitchen clean.

wipe up, put out, sanitize, wash, clean

1. You should learn how to ________ a fire with an extinguisher.

2. Pick up everything you drop, and ________ everything you spill.
3. I should always ________ my hands.
4. We have to ________ all the things before we leave the kitchen.
5. Everything in the kitchen should be ________ .

Activity 2

Task Here are some kitchen rules students should know. Could you add any more rules to them?

Kitchen supplies cannot be taken home.

Kitchen doors cannot be propped open but the doors on the North and South sides can be lifted for ventilation purposes.

Kitchen utensils cannot be borrowed.

Proper food preparation cannot be postponed.

Everyone is not allowed to wear shorts in the restaurant.

Sinks must be cleaned after use.

Dish sinks and their surrounding areas should be cleaned and wiped dry after use.

Clean clothing and close-toed shoes must be worn while you work in the kitchen.

Activity 3

Task 1 Read the following numbered English words about Kitchen Floor and try to write out their Chinese versions.

<table>
<tr><td>1. Freezer</td><td>2. Cold kitchen</td><td>3. Butchery</td><td>4. Pastry</td><td>5. Beverage cooler</td><td rowspan="3">13. Pick-up area</td></tr>
<tr><td rowspan="2">12. Kitchen store/
Pantry area</td><td colspan="4">6. Chef's office
7. Hot kitchen</td></tr>
<tr><td>8. Pot-washer</td><td>9. Vegetable preparation</td><td>10. Fish section</td><td>11. Scullery</td></tr>
</table>

1. ________________________________
2. ________________________________
3. ________________________________
4. ________________________________
5. ________________________________
6. ________________________________
7. ________________________________
8. ________________________________
9. ________________________________
10. ________________________________
11. ________________________________
12. ________________________________
13. ________________________________

Task 2 Further reading.

When finished with a dish or eating utensil, rinse it off before placing it in the sink. The simple act of rinsing can save much time when the dishes are washed.

After creating a horrible mess in a cooking pot or frying pan, clean the pot or pan before letting it dry. This will save hours of scrubbing work and damage to pans.

Do not place garbage in a trash receptacle unless there is a garbage bag in the receptacle.

Do not continue to place garbage into a trash receptacle after it is full. While this may seem to save work, it does not. When the trash receptacle is full, the garbage simply spills onto the floor where it has to be picked up again. This creates extra work and the garbage bag will have to be changed anyway.

Do not attempt to do the dishes. Cleaning dishes is a complex process involving soap, water, and several cleaning devices. It should only be performed by those familiar with the activity.

Use a rag to perform Steps 1 and 2. If the situation has progressed to the point where you feel you need to use another cleaning utensil, the utensil will surely be ruined in the process.

If you are unable to carry out Steps 1 to 6, please do not enter the kitchen under any circumstances. The house can be accessed via the front door and arrangements should be made for food to be left on the dining room table at appropriate times.

Unit 3 Verbs Used in Processing Pastry

Learning goals

To master verbs used in processing pastry.

To apply these verbs into sentences.

To use these verbs in daily dialogues.

Vocabulary

pastry [ˈpeɪstri] *n.* 面粉糕饼;馅饼皮

dough [dəu] *n.* 生面团

twist [twɪst] *v.* 缠绕;捻;搓;拧;扭曲

fold [fəʊld] *vt.* & *vi.* & *n.* 折叠;交叉; 折层;折痕

knead [niːd] *v.* 揉(面粉等);按摩;捏制

stretch [stretʃ] *v.* & *n.* & *adj.* 伸展;延伸; 可伸缩的

flatten [ˈflætn] *vt.* & *vi.* 使变平

braid [breɪd] *v.* 编织

rub [rʌb] *v.* 擦;揉;搓;

rolling pin 擀面杖

wrap [ræp] *v.* 包;裹;覆盖;

prick [prɪk] *v.* 刺;扎

crosshatch [ˈkrɒshætʃ] *vt.* & *n.* 画交叉排线

knock back 揉打

puff pastry 泡芙
garnish [ˈgɑːnɪʃ] *v.* & *n.* 装饰;配菜
demonstrate [ˈdemənstreɪt] *vt.* & *vi.* 示范;演示

Dialogue

(*A=Apprentice*, *C=Chef.*)

A: Good morning, Chef! What are we going to make today?

C: Morning. Today we are going to make a fruit cake.

A: What shall we do first?

C: Prepare the pastry first. Do you know how to work the pastry with a rolling pin?

A: Sorry, could you demonstrate once for me?

C: All right. Now let's begin. We should divide the dough and then flatten them.

Activity 1

Task 1 Try to write the names or actions in English below the pictures.

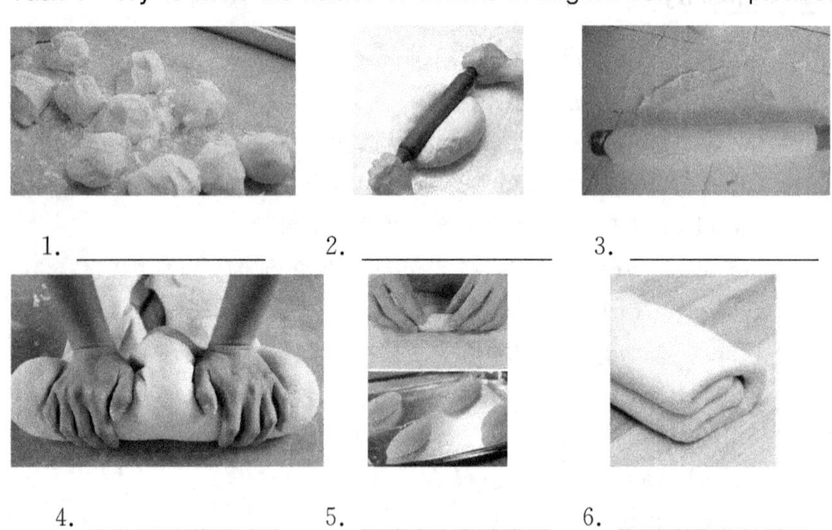

1. ________ 2. ________ 3. ________

4. ________ 5. ________ 6. ________

7. ______________ 8. ______________

Task 2 Read and try to remember the following expressions used in processing pastry.

处理面团工序

Work the pastry.
Knock back the pastry.
Twist the pastry.
Fold the pastry.
Braid the dough.
Knead the dough.
Rub the flour and butter together.
Stretch the pastry.
Roll out the pastry.
Flatten the pastry.
Let's flatten it with a rolling pin.
Unroll the pastry.
Shape the pastry.
Divide the dough.
Wrap it in pastry dough.
Prick the pastry with a fork.
Crosshatch the pastry.
Scrape up the dough. I am going to ice a cake.
Let's put a layer of chocolate icing on the cake.
Layer cream on the cake.
Line the mold.
Flour the cake pans.

I will spread the whipping cream, cherry filling, and frosting.

Put icing on the cake.

Smooth the cake.

Put some food coloring in the cream for the cake.

Decorate the edges.

Garnish the cake with whole cherries.

Loosen the cake.

Turn out the cake.

Make some puff pastries.

Activity 2

Task Translate the fallowing sentences into English.

1. 请用擀面杖把面团擀平。

____

2. 那个小男孩把面团揉成了球状。

____

3. 这个鱼就用黄瓜片做配菜。

____

4. 把面团展开然后分成3份。

____

5. 我想给蛋糕上撒一层糖霜。

____

6. 你能给我一些面粉吗？我想做泡芙。

____

Activity 3

Task Reading comprehension.

Direction: Read the following passage, work in pairs, fill in the blanks of the chef's job description and discuss with your partner.

Chef

The word "chef" is borrowed (and shortened) from the term "chef de cuisine"(French), the director or head of a kitchen. In English, the title "chef" in the culinary profession originated in the *haute cuisine* of the 19th century. Today it is often used to refer to any professional cook, regardless of rank, though in most classically defined kitchens, it refers to the head chef; others, in North American parlance, are "cooks".

Below are various titles given to those working in a professional kitchen and each can be considered a title for a type of chef. Many of the titles are based on the brigade de cuisine(or brigade system) documented by Auguste Escoffier, while others have a more general meaning depending on the individual kitchen.

Chef de cuisine, executive chef and head chef

This person is in charge of all things related to the kitchen, which usually includes menu creation, management of kitchen staff, ordering and purchasing of inventory, and plating design. Chef de cuisine is the traditional French term from which the English word chef is derived. Head chef is often used to designate someone with the same duties as an executive chef, but there is usually someone in charge of them, possibly making the larger executive decisions such as direction of menu, final authority in staff management decisions, etc. This is often the case for chefs with multiple restaurants.

Sous-chef

The sous-chef de cuisine (under-chef of the kitchen) is the second in command and direct assistant of the chef. This person may be responsible for scheduling and substituting when the chef is off-duty and will also fill in for or assist the chef de partie(line cook) when needed. This person is responsible for inventory, cleanliness of the kitchen, organization and constant training of all employees. The "sous-chef" is responsible for taking commands from the chef and following through with them. The "sous-chef" is responsible for line checks and rotation of all product. Smaller operations may not have a sous-chef, while larger operations may have several.

Chef de partie

A chef de partie, also known as a "station chef" or "line cook", is in charge of a particular area of production. In large kitchens, each station chef might have several cooks and/or assistants. In most kitchens, however, the station chef is the only worker in that department. Line cooks are often divided into a hierarchy of their own, starting with "first cook", then "second cook", and so on as needed.

Commis (Chef)

A commis is a basic chef in larger kitchens who works under a chef de partie to learn the station's responsibilities and operation. This may be a chef who has recently completed formal culinary training or is still undergoing training.

Station-chef titles which are part of the brigade system include:

English	French	Description
sauté chef	saucier	
fish chef	poissonnier	
roast chef	rôtisseur	
grill chef	grillardin	
fry chef	friturier	
vegetable chef	entremetier	
pantry chef	garde manger	
butcher	boucher	
pastry chef	pâtissier	

Unit 4 Verbs Used in Processing Fish and Meat

Learning goals

To know different terms of processing fish and meat.

To describe how to process fish and meat in English.

To be familiar with the expressions of processing fish and meat.

Vocabulary

rinse [rɪns] *n.* & *v.* 清洗；冲洗
scale [skeɪl] *n.* & *v.* 刻度；等级；刮鳞；剥落
gut [gʌt] *n.* & *vt.* 内脏；取出内脏
cut [kʌt] *n.* & *v.* 切；割；剪；切口；
cut off 切断；停止运行
cut open 剖；切开
flake [fleɪk] *n.* & *v.* 薄片；剥落
bone [bəʊn] *n.* & *v.* 骨头；除去骨头
decorate [ˈdekəreɪt] *v.* 装饰；布置；装修
lard [lɑːd] *n.* & *v.* 猪油点缀；涂油于；润
tie up 扎紧

Dialogue

(*C1=Commis*, *C2=Chef.*)
C1：What are you doing?
C2：I'm beating meat with a cutlet bat.
C1：Why?
C2：This veal is not flat.
C1：So you are using a cutlet bat.
C2：That's right.
C1：Shall we need to bone the veal before flattening it as well?
C2：Certainly, or we could not flatten it.
C1：All right. I know how to deal with it now.
C2：Good, do you know how to deal with fish?
C1：Er…I only know we should scale it.
C2：Yes! Scale it with a scaler and then gut it.
C1：OK, I've got it now.

Activity 1

Task 1 Try to write the names or actions in English below the pictures.

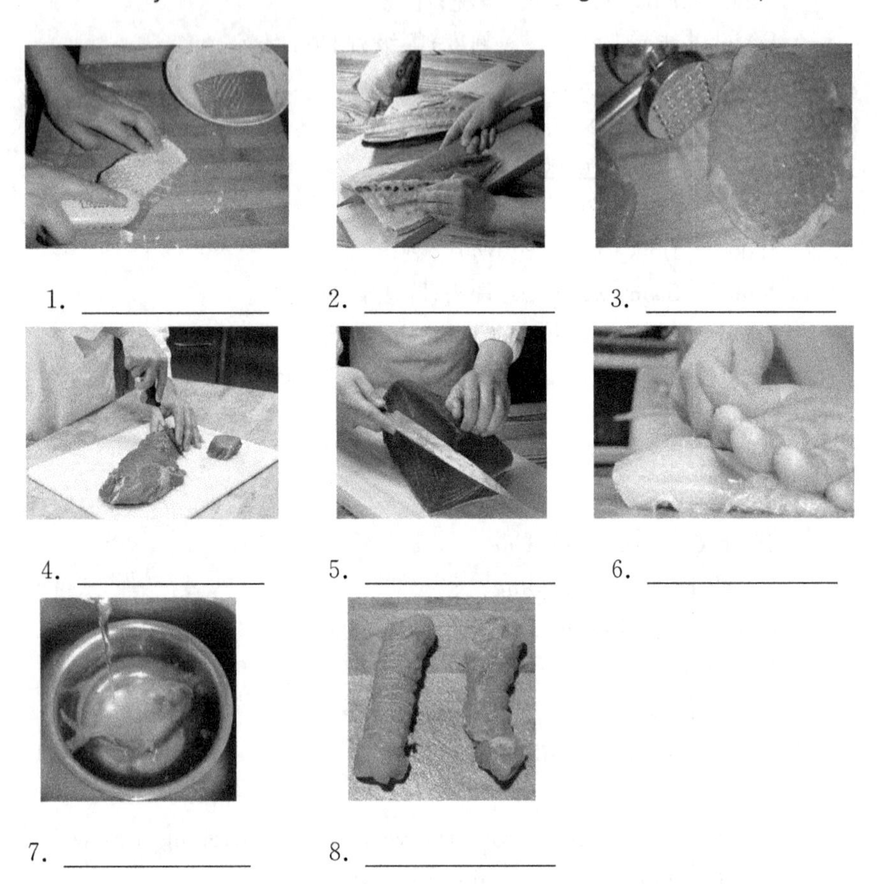

1. ____________ 2. ____________ 3. ____________

4. ____________ 5. ____________ 6. ____________

7. ____________ 8. ____________

Task 2 Read and try to remember the following expressions used in processing meat and fish.

1. 处理鱼类工序

Rinse the fish.

Cut off the fins.

Scale it. Then gut it.

Cut open the stomach of the fish. Then take out the guts.

Flake the fish.

Flatten the escalopes.

2. 处理肉类工序

Bone the beef.

Lard the meat.

Decorate the chicken.

Tie up the meat.

Flatten the meat.

Beat the meat.

Activity 2

Task 1 Translate the following sentences into English.

1. 烧鱼之前先去掉鱼鳞。

____

2. 请把牛肉去骨，并把肉拍平。

____

3. 把鱼肚子切开，然后取出内脏。

____

Task 2 Put the following sentences into the correct order according to the time sequence.

1. Cook the fish.
2. Gut the fish.
3. Rinse the fish.
4. Scale the fish.
5. Pan-fry the fish.

________ ________ ________ ________ ________

Task 3 Reading comprehension.

Directions: Read the following passage and retell the process of making beef steaks.

How to Make Beef Steaks

Step 1: Getting your ingredients ready

4 steaks (rump or sirloin)

4 tablespoons of crushed peppercorns

3 tablespoons of olive oil

3 tablespoons of unsalted butter

3 chopped shallots

30 ml of cognac

60 ml of thick cream

salt and pepper

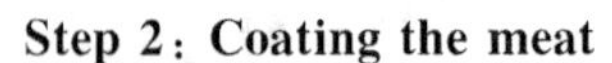

Step 2: Coating the meat

Spread the peppercorns evenly on a plate and press the steaks firmly down on both sides. For extra flavour, do this for an hour or two before you are ready to cook.

Step 3: Seasoning

Season the steaks with a little salt and pepper and thoroughly cover them in oil.

Step 4: Cooking the steaks

Heat the oil over moderate to high heat and add a knob of butter. When the oil is hot, add the steaks and sear on both sides. Depending on the size, cooking for about three minutes will leave them medium rare. Don't turn the steaks too much, so as not to dislodge the peppercorns. When cooked, take the steaks out of the skillet.

Step 5: Sautéing the shallots

Pour out the excess oil from the skillet and add a knob of butter. Sauté the shallots until they turn soft and brown.

Step 6: Making the sauce

When the shallots are cooked, add the cognac. Allow the butter/cognac mixture to reduce to about half and add the cream and green peppercorns. Cook over a high heat for about a minute, stirring well until reduced and season it with salt.

Step 7: Serving

Place the steaks on serving plates and pour the sauce over the top.

Unit 5 Verbs Used in Processing Other Food

Learning goals

To know different terms of processing other food.

To describe how to process other food in English.

To be familiar with more expressions of processing other food.

Vocabulary

wring out *vt.* 绞出;拧掉;扭干

refrigerate [rɪ'frɪdʒəreɪt] *v.* 使冷冻;使冷藏

soak [səʊk] *n.* & *v.* 浸;浸泡;渗透

cube [kjuːb] *n.* & *v.* & *adj.* 立方;求……的立方;立方的

diagonally [daɪ'æɡənəli] *adv.* 斜对地;对角地

mince [mɪns] *v.* & *n.* 切碎;肉酱

dice [daɪs] *n.* & *v.* 骰子;小立方;把……切成丁

trim off 修剪

peel [piːl] *n.* & *v.* 果皮;削皮

remove seeds 去籽

grate [ɡreɪt] *v.* 摩擦;磨碎

crush [krʌʃ] *n.* & *v.* 压碎;压榨

grind [ɡraɪnd] *v.* 磨;压迫;碾碎

mash up 捣烂

toss [tɒs] *n.* & *v.* 投掷;摇荡

squeeze [skwiːz] *n.* & *v.* 挤压;塞进;压榨

grease [ɡriːs] *n.* & *vt.* 油脂;用油脂涂;上油

stuff [stʌf] *n.* & *vt.* 塞满;填满

marinate ['mærɪneɪt] *v.* (做沙拉时将肉鱼等在调味汁中)浸泡;腌制

serrate [ˈsereɪt] *v.* & *adj.* 使成锯齿状；锯齿状的
strain [streɪn] *n.* & *v.* 拉紧；紧张
sift [sɪft] *v.* 筛
skim [skɪm] *n.* & *v.* 浮沫；撇去；掠过
sauté [ˈsəʊteɪ] *n.* & *v.* & *adj.* 嫩煎菜肴；快炒；煎的；炒的

Dialogue

(*C1=Commis*, *C2=Chef.*)

C1: What shall I do with the dishes?
C2: Rinse the vegetables first, and then cut the cucumbers into cubes.
C1: OK, I have finished.
C2: Beat the eggs with a whisk and grind the garlics.
C1: Shall I remove the seeds of the peppers as well?
C2: Right. Go ahead.
C1: All right. What are we going to do with the potatoes?
C2: Mash them up.
C1: Shall we boil them first?
C2: Yes, but we should peel them before we do that.
C1: OK.

Activity 1

Task 1 Try to write the actions in English below the pictures.

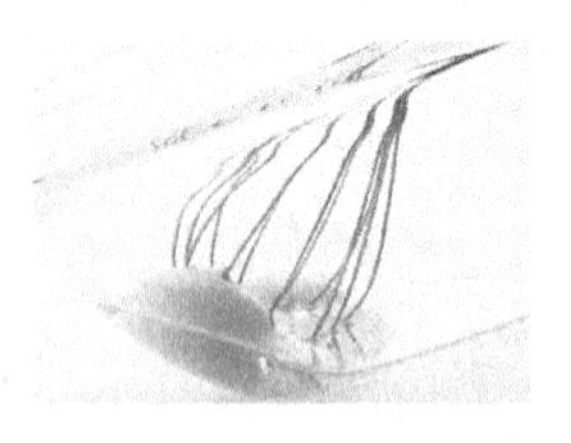

1. ______________ 2. ______________ 3. ______________

Task 2 Read and try to remember the following expressions used in processing food.

1. 拧	Wring out the dishcloth.
2. 擦干	Dry the dish with a towel.
3. 拿过来	Bring me some apples, please.
	Get me some lemons.
4. 准备	Fix the sauce.

5. 做	Make the waffles, please.
	Make puree with vegetables.
6. 称重	Please weigh the limes for me.
7. 备料	Prepare some cheese for the cheese board, please.
8. 冷藏	Refrigerate them.
9. 用力搓洗	Should I scrub the potatoes?
10. 清洗	Wash the dishes with detergent.
	Clean the vegetables.
11. 浸泡	Soak the cauliflower in salted water.
	Steep the bread in milk for a few minutes.
12. 切	Carve the roast, please.
13. 切块	Cut it into cubes.
14. 切一半	Cut it into halves.
15. 切成三等分	Cut it into thirds.
16. 切成四等分	Cut the onion into fourths.
	Cut the apple into quarters.
17. 切成五等分	Cut the carrot into fifths.
18. 切成十等分	Cut the cake into ten portions.
19. 对角切	Cut the sandwich diagonally.
20. 沿边切	Line the mold.
21. 剖开	Open the coconut.
22. 切块	Cut it into pieces.
23. 切成条	Please cut the beef into strips.
24. 切丝	Shred the cheese.
25. 切片	Please slice the aubergines.
26. 切成细末	Should I chop the onion fine?
27. 切碎	I need to cut up some onions.
	Mince the beef.
28. 切丁	Should I dice the carrots?
29. 切开	Should I split them down the middle?
30. 切除	Should I trim off the ends?
	Cut up the bones.

31. 剥(削)皮	Peel ten cloves of garlic.
	Peel an apple.
32. 去籽	Please remove the seeds.
	Seed the pumpkin.
33. 磨碎	Grate the cheese.
	Please crush the garlic!
	Please grind the garlic.
34. 捣成泥	Mash up the bananas.
35. 剁碎	Chop up the fennel.
36. 刮丝	Serrate the lemon.
37. 用手撕开	Tear the lettuce into bite-size pieces.
38. 搅拌速度	Blend at high/medium/low speed.
39. 搅拌	I am going to stir some ingredients.
	Toss the salad with dressing.
40. 搅拌频率	Stir constantly.
	Stir occasionally.
41. 混合	Please mix the garlic with butter.
42. 填充	Please stuff the peppers.
	We will fill the bottoms with peas and carrots.
43. 挤压	Squeeze the lemons.
44. 打蛋	Beat the eggs.
45. 拍打肉	Beat the meat with a cutlet bat.
	I'm flattening the veal.
46. 做薄肉片	I'm making some escalopes.
47. 沥干	To drain them.
	Drain the carrots in a colander.
48. 过滤	Strain sauces.
49. 去除表面杂质	Skim the liquid.
	Skim the hot fat.
50. 过筛	Sift the breadcrumbs.
	Sift the flour, please.
	Sift the icing sugar, please.

51. 加入	Please add salt to the garlic.
	Put the peaches into the fruit salad.
52. 刷上奶油	Brush the bread.
53. 抹奶油	Butter the mushrooms.
54. 加盐	Salt the cauliflower.
	Salt the pasta water.
55. 涂上油脂	Baste the leg of lamb.
56. 涂抹	Coat the chicken with the mixture.
	Spread the jam.
57. 腌渍	I want to marinate some chicken.
	I want you to soak the beef in wine.
58. 剥壳	It's for opening oyster shells. Oysters are a kind of shellfish.
59. 插入肉类温度计	Stick a meat thermometer in the meat.
60. 测肉温度	Test the meat, please.
	Test the roast with a meat thermometer.
61. 融化	Melt some butter.
62. 揉	Rub the flour and butter together.
63. 加热	Heat some cooking oil in it.
64. 预热	Preheat the oven to three hundred and fifty degrees.
65. 重新加热	Reheat the meat.
66. 加热	Heat the meat.
67. 降温	Cool the meat.
68. 抹油	Grease two cake pans.
69. 刺孔	Prick the orange.
70. 做烤肉串	Should we make brochettes today?
71. 插入叉子	Stick a roasting fork into the meat.
72. 拔出叉子	Pull the roasting fork out of the meat.
73. 加水	Pour some water into the stockpot. Do not fill it to the top.
74. 装满	Fill up the bottle.
75. 冷藏	Refrigerate them.

76. 磨刀	I am going to sharpen my knife.
77. 水煮	Please boil the corn.
	Please poach the artichokes.
78. 煮沸	Bring the water to a boil.
79. 烤	Bake the cheese pies for fifteen minutes.
	Sear the meat in hot fat.
	I will grill the brochettes on the grill.
80. 烤面包	Toast the bread.
81. 油炸	Fry the chicken.
82. 放入油炸篮	Put the frying basket in the deep fryer.
83. 炖	I will simmer the stew over a low flame.
	Let the sauce simmer till the wine has evaporated.
	Please braise the beef.
84. 蒸	Steam the rice.
85. 煎	Fry the crepes.
86. 嫩煎	Sauté the peas.
87. 做酱汁	I am going to make a sauce.
88. 做清汤	Make some consomme.
89. 上酱汁	I am going to serve sauce.
90. 卷起	Now I will fold the omelet.
91. 刷上	Brush the vegetables with butter.
92. 取出	Lift the roast out of the pan.
93. 倒	Pour onion soup on top.
	Pour the ingredients into the baking pans.
94. 倒空	Empty a can of tomato puree.
95. 冷却	Let the cream sauce cool slightly.
96. 放到锅上	Put the roast in a pan.
97. 放到中央	Place the tuna in the center of the salad.
98. 从锅中拿出	Take the meat out of the pan.
99. 从锅中倒出	Slide the omelet onto the plate.
100. 移开	Take the pan off the fire.
101. 舀	Scoop up the ice cream.

102. 围绕	Surround the dish with meatballs.
103. 用汤匙淋	Spoon sauce over the stuffed tomatoes.
104. 淋上佐料	Put some cream on the blackberries.
105. 调味	Season the soup before serving.
106. 撒面包屑	Bread the escalope.
107. 洒上	Sprinkle some cheese on the pizzas. Sprinkle the eggs with processed cheese.
108. 移动	I am going to transfer some vegetables to a plate. Transfer the peas from the colander to the pot.
109. 递送	Hand the chef a roasting fork.
110. 上菜	Serve the potatoes.
111. 盖起来	Cover the food.

Activity 2

Task 1 Translate the following sentences into English.

1. 使劲擦洗一下那地板。

____

2. 瘦肉剁碎,加调味捞匀。

____

3. 请把这个木瓜去籽。

____

4. 玛丽让丈夫帮她把烤箱预热到180度。

____

5. 他把醋洒在鱼和薯条上。

____

6. 服务员往我的饮料中挤了少量柠檬汁。

____

7. 不能磨得比这个再细了。

____

Task 2 Match the verbs on the left with the phrases on the right.

1. squeeze　　a. the mixture with a wooden spoon

2. melt	b. the potatoes and boil in a pan
3. beat	c. the cheese and add to the sauce
4. mix	d. the sauce over the meat and serve
5. chop	e. the ham as thin as possible
6. stir	f. the eggs until they turn light and fluffy
7. grate	g. a lemon over the fish
8. slice	h. a little butter in a frying pan
9. pour	i. the vegetables into small pieces
10. peel	j. all the ingredients together

Chapter 2

Kitchenware

Unit 1 Frying Baskets, Skimmers, Colanders, Sieves, Conical Strainers and Chinois

Learning goals

To master the names of the following kitchenware: frying basket, skimmer, sieve, conical strainers and chinois.

To know the usage of the above kitchenware in the kitchen.

To apply these names of kitchenware into dialogues.

Vocabulary

deep fryer 油炸锅
filter [ˈfɪltə(r)] *n.* & *v.* 过滤;渗透;过滤器
skimmer [ˈskɪmə] *n.* 撇取浮物的器具;网勺
frying basket 油炸篮
colander [ˈkʌləndə(r)] *n.* 滤器;漏勺
conical strainer 圆锥形过滤器
chinois [ʃiːˈnwɑː] *n.* (厨房用)漏勺
particle [ˈpɑːtɪkl] *n.* 颗粒;微粒

Dialogue

(*C1=Commis*, *C2=Chef.*)

C1: How shall I cook these potato chips?

C2: Put the potato chips in a frying basket.

C1: And then what?

C2: Put the frying basket in the deep fryer.

C1: Now?

C2: Yes, now. We are making British fries.

C1: What shall I do with the carrots and the peas?

C2: Are they cooked?

C1: Yes.

C2: Put the carrots in a colander and put the peas into the conical strainer.

C1: Why?

C2: To drain them of course.

C1: After I drain them?

C2: Butter and salt them.

C1: While they are in the colander and the conical strainer?

C2: No. After you take them out.

Activity 1

Task 1 Try to write the names in English below the pictures.

1. ______________ 2. ______________ 3. ______________

4. ______________ 5. ______________ 6. ______________

Task 2 Read and try to remember the following expressions about the kitchenware.

1. 油很脏	The oil in the deep fryer is dirty. The top layer of the oil now has a lot of small pieces of floating food.
2. 询问是否使用滤网	Did you clean the oil with a filter?
3. 告知沥油时间	Yes, about twenty minutes ago.
4. 使用漏勺	Yes, I see. Use a skimmer, please.
5. 吩咐拿/使用漏勺	Could you hand me a skimmer? Use a skimmer, please.
6. 漏勺用处	It is used to remove food particles from oil.
7. 放进油炸篮	Put the potato chips in a frying basket.
8. 放进油炸锅	Put the frying basket in the deep fryer.
9. 放进滤锅	Put them in a colander.
10. 询问哪一个滤碗	Which colander?
11. 指定滤碗	A colander with big holes.
12. 询问筛网用处	What is a sieve for? Should I sift this flour with a sieve?
13. 筛网用处	It is for flour or breadcrumbs or icing sugar.
14. 请人用筛网	Use a sieve, please.
15. 请人用圆锥形滤网	Use a conical strainer.
16. 吩咐拿圆锥形滤网	Hand me a conical strainer, please. I need a conical strainer.
17. 没有圆锥形滤网	I don't have a conical strainer.
18. 告知没关系	That's all right. I'll find one.

19. 吩咐拿锥形漏勺　Give me a chinois, please.
20. 询问用处　What will you strain?
21. 告知锥形漏勺用处　A sauce.

Activity 2

Task 1　Translate the following sentences into English.

1. 油炸锅里油的表层现在有很多悬浮食物残渣。

____

2. 我要用筛网来筛这些面粉吗?

____

3. 请问你用圆锥形滤网来干吗?

____

4. 我要用漏勺把油表层清干净。

____

5. 我需要一个锥形漏勺,能麻烦你递给我吗?

____

Task 2　Do the task in pairs. Look at the following pictures and figure out their names in English. Then try to tell what they are used for.

1.  ______________

Description:

____

____

____

2.

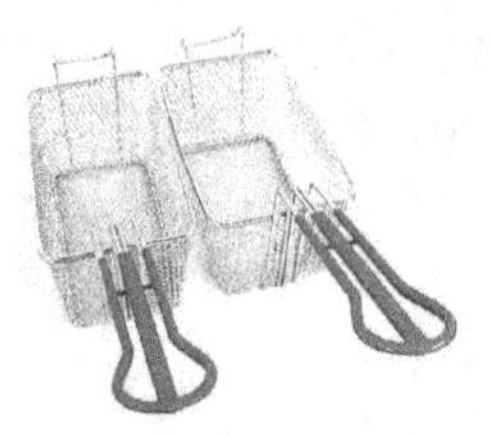

____

Description(usage):

____

____

____

Activity 3

Task Further reading.

Colander

A colander is a bowl-shaped kitchen utensil with holes in it. It is used for draining food such as pasta or rice. The perforated nature of the colander allows liquids to drain through while retaining solids inside. It is sometimes also called a pasta strainer or kitchen sieve. Conventionally, colanders are made of a light metal, such as aluminium or thinly rolled stainless steel. Colanders are also made of plastic, silicone, ceramic, and enamelware. The word colander comes from the Latin "colum" meaning "sieve".

Unit 2 Forks and Scissors

Learning goals

To master the names of forks and scissors in the kitchen.

To know the usage of the forks and scissors.

To apply the names of forks and scissors in expressions and dialogues.

Vocabulary

fork [fɔːk] *n.* & *v.* 叉子;使……成叉形;用叉叉起
skewer ['skjuːə(r)] *n.* & *v.* 串肉扦;用扦串肉
shish kebab *n.* 羊肉串
knife set 套刀
oyster knife 牡蛎刀
boning knife 去骨刀
chopper ['tʃɒpə(r)] *n.* 斧头;砍刀
grapefruit knife 葡萄柚刀
carving knife 雕刻刀
cheese knife 起司刀
fish scissors 鱼剪刀
poultry ['pəʊltri] *n.* 家禽;家禽肉
poultry shear 家禽剪

Dialogue

(*C1=Commis*, *C2=Chef*.)
C1: Shall I make brochettes today?
C2: Yes. Let's make shish kebabs.
C1: OK, so I have to use some skewers.
C2: Yes. And we have to make some fruit salads and some seafood dishes too.
C1: So I also need a pair of fish scissors, an oyster knife, a peeler and a grapefruit knife as well.
C2: Of course.

Activity 1

Task 1　Try to write the names in English below the pictures.

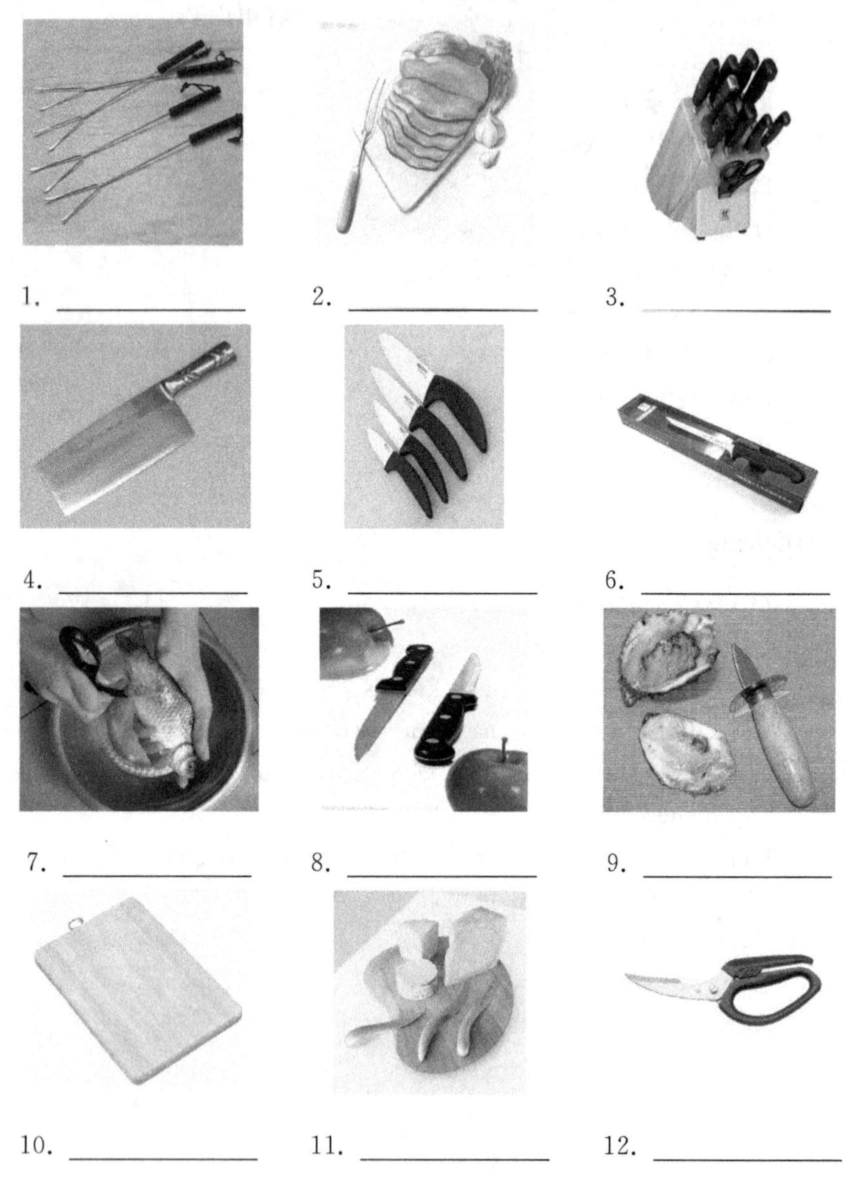

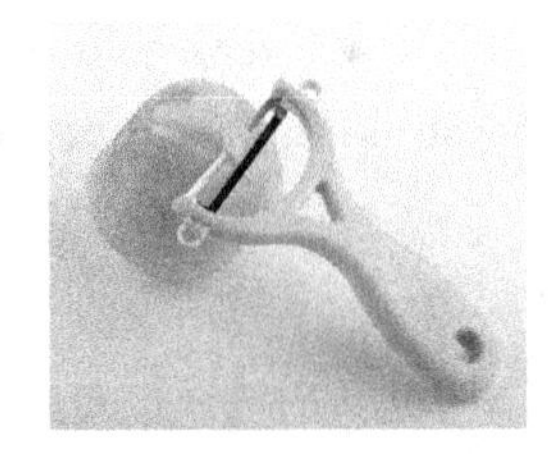

13. ______________

Task 2 Read and try to remember the following expressions about the kitchenware.

1. 吩咐用烤叉	Lift the meat with the roasting fork.
2. 吩咐拿烤肉叉	Give me some skewers，please.
3. 烤制串肉	I will put the meat on the skewers. Use skewers for the shish kebab.
4. 询问物品为何物	What are these?
5. 告知为刀具	They are kitchen cutters.
6. 就刀具进行询问	Is my cook's knife for special jobs? What are these knives for? What else can I use kitchen cutters for?
7. 刀具用处	Your cook's knife is for many different things. You can cut many different things with them. Kitchen cutters are very useful. This is a boning knife. A boning knife is used to remove bones from meat. It's an oyster knife. It's for opening oyster shells. Oysters are a kind of shellfish.
8. 询问可否用刀切肉	Can I cut up this meat with my cook's knife?
9. 询问可否用砍刀砍牛肉	Can I use a chopper to chop some beef?
10. 询问肉是否带骨	Are there any bones in the meat?
11. 可用该刀切肉	Sure. Use your cook's knife to cut the meat.
12. 不可用该刀切肉	No. Use a cleaver.
13. 询问适用刀具	What kind of knife should I use?
14. 告知适合用刀	Use your cook's knife.

	Use a carving knife. I will give you one.
	Use this big cheese knife with a double handle.
	Use a grapefruit knife.
15. 刀具无法使用	My knife is not cutting this beef.
16. 询问可否在桌上切肉	Can I chop up the beef here on the counter?
17. 在砧板上切	No. Chop it on the chopping block.
18. 询问可否用刀具去鱼鳞	Can I scale it with this knife?
19. 用鱼剪取出内脏	Yes. And use these fish scissors to gut it.
20. 告知剪刀用途	These are special scissors. Use them to cut open the stomach of the fish.
21. 询问是否为剪刀	Are these scissors?
22. 告知为家禽剪	No. They are poultry shears.
23. 询问用途	— You mean they are for cutting up birds? — Yes. Especially chicken.

Activity 2

Task 1 Translate the following sentences into English.

1. 请问这把奇特小刀的用途是什么?

____

2. 我可以用厨房刀具来打开这瓶酸辣酱的盖子吗?

____

3. 去骨刀的用途是剔骨。

____

4. 你可以削个苹果给我吗?

____

5. 一套刀具包含了几种不同用途的刀?

____

6. 普通剪刀和家禽剪有什么不一样?

____

Task 2 Different knives and scissors are used for preparing different food (meat, fish, vegetables, fruit, etc.). Match the following knives or scissors in Column A with the food in Column B.

Column A	Column B
pallet knife	chicken
grapefruit knife	beef
fish scissors	cake
cleaver	bone
cheese knife	orange
bone saw	fish
poultry shears	cheese
oyster knife	meat
steak knife	oyster

Activity 3

Task Reading Comprehension.

Directions: Work in pairs, read the passage and discuss with your partner about the differences between ceramic knives and the other knives.

Ceramic Knives

A ceramic knife is a knife made out of very hard and tough ceramic, often zirconium dioxide (ZrO_2; also known as zirconia). These knives are usually produced by dry pressing zirconia powder and firing them through solid-state sintering. The resultant blade is sharpened by grinding the edges with a diamond-dust-coated grinding wheel. Zirconia ranks 8.5 on the Mohs scale of mineral hardness, compared to 7.5 to 8 for hardened steel, and 10 for diamond. This very hard edge rarely needs sharpening.

Ceramic knives will not corrode in harsh environments, are non-magnetic, and do not conduct electricity. Because of their resistance to strong acid and caustic substances, and their ability to retain a cutting edge longer than forged metal knives, ceramic knives are a much more suitable culinary tool for slicing boneless meat, vegetables, fruit and bread. Since they are brittle they may break if dropped on a hard surface, and cannot be used for chopping through bones, or frozen foods, or in other applications which require prying, which may result in chipping or catastrophic failure. Several brands now offer a black colored blade made through an additional hot isostatic pressing (HIP) step, which improves the toughness.

Ceramic knives may present a security problem as ceramics are not seen by conventional metal detectors. To hinder misuse of concealed knives, many manufacturers include some metal to ensure that they are seen by standard equipment. Ceramic knives may be detected by extremely high frequency scanners (e. g. millimeter wave scanners) and X-ray backscatter scanners.

Unit 3 Spoons, Spatulas, Ladles, Bowls and Cutlet Bats

Learning goals

To master the names of spoons, spatulas, ladles bowls and cutlet bats in the kitchen.

To know the usage of the spoons, spatulas, ladles, bowls and cutlet bats.

To apply the names of spoons, spatulas, ladles, bowls and cutlet bats in expressions and dialogues.

Vocabulary

palette knife 调色刀
layer [ˈleɪə] *n.* & *v.* 层;分层
spatula [ˈspætʃələ] *n.* 铲;抹刀;压舌片
ladle [ˈleɪdl] *n.* & *v.* 勺子
slotted spoon 有孔汤匙
basting brush 涂油刷
mixing bowl 搅拌碗
soup tureen 盛汤盖碗

cutlet bat 把肉拍平的木板或木槌

mallet [ˈmælɪt] *n.* 木槌

escalope [ˈeskəlɒp] *n.* 薄肉片;(裹面包屑和鸡蛋的)炸肉块

Dialogue

(*C1*=*Commis*, *C2*=*Chef*.)

C1: What do you want me to do today?

C2: Oh, such a lot of things we should finish today. Stir the ingredients in a bowl now.

C1: With what shall I stir these ingredients?

C2: Use a spatula.

C1: A wooden spatula?

C2: Sure. And after that you could butter the vegetables with a brush.

C1: What exactly should I do?

C2: First, put the brush in hot butter. Then brush the vegetables.

C1: Right.

C2: And then you should serve the sauce and the soup.

C1: With what?

C2: Use this small ladle to serve the sauce and take the big one to serve the soup.

C1: OK.

Activity 1

Task 1 Try to write the names in English below the pictures.

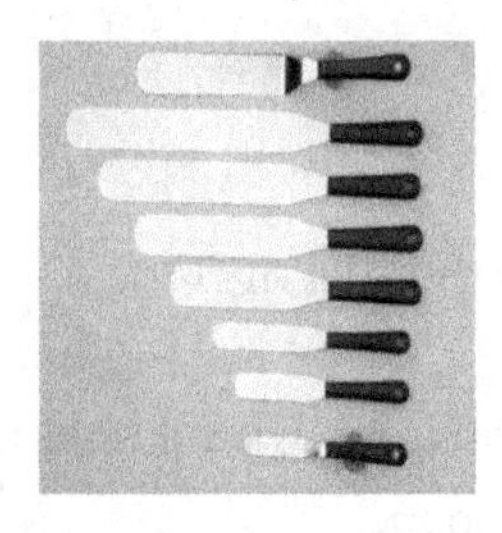

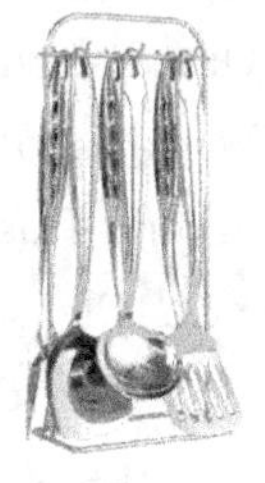

1. ______________ 2. ______________ 3. ______________

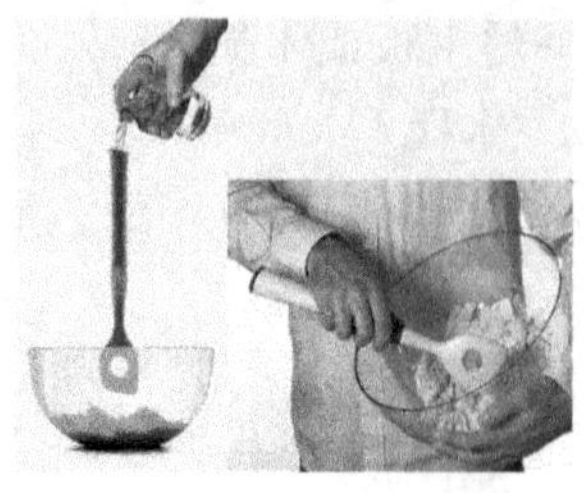

4. ______________ 5. ______________ 6. ______________

7. ______________ 8. ______________

Task 2 Read and try to remember the following expressions about the kitchenware.

1. 抹糖霜	Let's put a layer of chocolate icing on the cake.
2. 询问用什么工具	With what?
3. 用抹刀	With a palette knife.
	Use this palette knife.
4. 搅拌材料	Stir the ingredients in the bowl.
5. 询问用什么工具搅拌	With what should I stir the ingredients?
6. 告知使用的各类汤匙	Use a wooden spatula.
	Use a spoon.
	Use a serving spoon.
7. 询问哪一种汤匙	Which spoon?
8. 询问是否用大分菜匙	Should I use a big serving spoon?
9. 表示没错	Yes. Please do it.
10. 用有孔汤匙	Use a slotted spoon, one with a hole in it.
	Use this slotted spoon.
11. 使用原因	It is very good for stirring.

12. 询问可否使用漏眼匙	Can I serve peas with a slotted spoon?
13. 同意使用有孔汤匙	I'll use a slotted spoon to stir the ingredients.
14. 告知称呼	It's a serving spoon.
	We call it a serving spoon.
15. 请人拿长柄勺	Give me a ladle, please.
	I need a big ladle. I don't have one.
16. 询问用处	To serve the sauce?
	To serve the soup?
17. 告知大小汤勺	Take my big ladle.
	Take this small ladle.
18. 给蔬菜刷上奶油	I want to butter these vegetables.
19. 告知使用刷具	Use that basting brush.
20. 告知工序	First, put the brush in hot butter. Then, brush the vegetables.
21. 询问是否为搅拌碗	Is this a mixing bowl?
22. 告知没错	Yes, it is.
23. 允许使用	Sure. Go ahead.
24. 汤煮好	The soup is ready.
25. 询问菜肴盛放处	Where shall I transfer the stew?
26. 把汤端出去	The waiter will take the soup tureen now.
27. 吩咐用拍肉板打肉	Beat the meat with a culet bat.
28. 肉不平坦	This veal is not flat.
29. 使用拍肉板原因	So you are using a cutlet bat.
30. 告知为木槌	It's a mallet.
31. 木槌功用	Use it to flatten meat.
32. 询问是否看见原材料	Do you see those escalopes?
33. 吩咐用木槌打平肉片	Beat those escalopes with the mallet.

Activity 2

Task 1　Translate the following sentences into English.

1. 那是个很不错的生日蛋糕。接下来我们要做什么?

____

2. 请用抹刀在蛋糕上抹上一层糖霜?

____

3. — 这个汤匙叫什么?

— ______________________________

— 叫漏眼勺。

— ______________________________

4. 我们把它盛到汤碗里吧。

____

5. 请用木槌把肉拍成薄片。

____

6. 我可以在这个搅拌碗里拌蛋黄酱和金枪鱼吗?

____

Task 2　Read the following English words in Column A and try to match them with the Chinese versions in Column B.

1. brush	a. 抹刀
2. mallet	b. 木槌
3. spatula	c. 木勺
4. ladle	d. 拍肉板
5. wooden spoon	e. 刷子
6. pallet knife	f. 铲
7. cutlet bat	g. 长柄勺

Activity 3

Task　Reading comprehension.

Directions: Work in pairs, read the passage and discuss with your partner about a spatula.

Spatula

In kitchen utensils, a *spatula* is a tool with a broad flat blade used for mixing and spreading things, especially in cooking and painting. One variety is alternately named turner, and is used to lift and turn food items during cooking, such as pancakes and fillets. These are usually made of plastic, with a wooden or plastic handle to insulate them from heat. A frosting spatula is also known as a palette knife and is usually made of metal or plastic. Bowl and plate scrapers are sometimes called spatulas.

In British English a *spatula* is similar in shape to a palette knife without holes in the blade. A wide-bladed utensil with long holes in the blade used for turning food is a fish slicer.

Unit 4 All Kinds of Pans

Learning goals

To master the names of all kinds of pans in the kitchen.

To know the usage of all kinds of pans.

To apply all these names in expressions and dialogues.

Vocabulary

frying pan *n.* 煎锅;长柄平锅

crepe [kreɪp] *n.* 薄煎饼

sauté pan 炒锅

bain marie 保温锅

stockpot [ˈstɒkpɒt] *n.* 锅子

saucepan [ˈsɔːspən] *n.* 长柄深锅

consommé pan 清炖肉汤锅

stew pan 炖锅

simmer [ˈsɪmə(r)] *n.* & *v.* 炖;煨

flame [fleɪm] *n.* & *v.* 火焰;热情;燃烧

spider [ˈspaɪdə(r)] *n.* 长柄平底锅
oven [ˈʌvn] *n.* 烤箱;烤炉
braising pan 炖(肉)锅
fish kettle *n.* 煮鱼锅
roasting tray 烤盘

Dialogue

(*C1=Commis, C2=Chef.*)

C1: What's this?
C2: It's a consommé pan.
C1: What shall I do with it?
C2: Do you see the ingredients in this pan?
C1: Yes, I do.
C2: Stir the ingredients. I mean, mix the things in the consommé pan.
C1: What do you need then?
C2: I need a stew pan.
C1: Right now?
C2: Yes, now! We are going to make the stew.
C1: OK, shall we make the stew by simmering?
C2: Yes.

Activity 1

Task 1 Try to write the following names in English below the pictures.

1. ______________ 2. ______________ 3. ______________

4. ______________ 5.  ______________ 6. ______________

7. ______________ 8. ______________

Task 2 Read and remember the following expressions about the kitchenware.

1. 询问是否为煎锅	Is this a frying pan?
2. 告知为特殊的煎锅	Yes. But it is a special type of frying pan.
	— What is this frying pan used for?
	— It is a small frying pan for crepes.
3. 询问锅具名称	What do you call this pan?
4. 告知锅具名称	It's a sauté pan.
5. 询问盛菜处	Where shall I transfer the stew?
6. 询问菜是否在锅中	Is the stew in the pot?
7. 告知盛放处	Put it in a bain marie.
8. 保温锅功能	A bain marie will keep it hot.
9. 询问用哪一个锅	In what pan should I cook this chicken?
10. 告知使用长柄煮锅	Cook the chicken in a saucepan.
	Cook the chicken in butter in a saucepan.
11. 准备酱汁	Fix the sauce to go with the chicken.
12. 询问是否用同一个锅	In the same pan?
13. 表示肯定	Right.

14. 询问放置处 Where should I put the chickens?
15. 告知放置处 Put all three chickens in the stockpot.
16. 汤锅加水 Pour some water into the stockpot. Do not fill it to the top.
17. 告知火候 The water is simmering.
18. 询问这是什么 What's this?
19. 告知是高汤炖锅 It's a consommé pan.
20. 询问要用来做什么 What should I do with it?
21. 询问需要什么 What do you need?
22. 告知需要长柄炖锅 I want a stew pan.
23. 询问如何炖菜 How will you cook the stew?
24. 询问是否用小火 Will you use a low flame?
25. 告知以小火慢炖 I will simmer the stew over a low flame.
26. 告知要将菜移至其他容器 I want to transfer a lot of beets from this big pot.
27. 提供平底锅 Here. Use this spider.
28. 表示平底锅很大 This spider is very big.
29. 使用大平底锅的理由 You need a big spider to do the work quickly.
30. 询问如何烧牛肉 How do you braise beef?
31. 使用热油 Sear the meat in hot fat.
32. 使用烤箱 Cook it in the oven.
33. 使用焖锅 Cook it in a braising pan.
34. 询问是否为煮鱼锅 Is this a fish kettle?
35. 告知放鱼进去 Put the salmon in it.
36. 告知小火慢炖鱼 Simmer the fish slowly.
37. 询问烤盘用处 What should I do with this roasting tray?
38. 告知热油 Heat some cooking oil in it.
39. 告知欲做餐点 We will make crown roast.
40. 询问是否要放进烤箱 Should I put the roasting tray in the oven?
41. 告知锅具很烫 Watch out! This pressure cooker is hot.
42. 叮咛小心打开锅具 Lift the steam valve weight slowly, and check the inside of the cooker.

Activity 2

Task 1 Translate the following sentences into English.

1. 你可以用拍肉板把肉拍平吗?

____

2. 保温锅是用来保温菜的。

____

3. 小心！这个高压锅很烫。

____

4. 请问我该用这个烤盘做什么?

____

5. 请用长柄煮锅炖鸡肉。

____

6. 绝对不要将锅装到最满,当你要打开锅盖的时候,要非常慢并仔细地打开。

____

Task 2 Pairs work. Do the task in pairs. Look at the following pictures and figure out their names in English. Then try to tell what they are used for.

1. ______________

Description (usage):

____

____

____

2. ______________

Description (usage):

____

____

____

3.  ______________

Description (usage):

____

____

____

Activity 3

Task Reading comprehension.

Frying Pan

A frying pan, frypan, or skillet is a flat-bottomed pan used for frying, searing, and browning foods. It is typically 200 to 300 mm (8 to 12 in) in diameter with relatively low sides that flare outwards, a long handle, and no lid. Larger pans may have a small grab handle opposite the main handle. A pan of similar dimensions, but with vertical sides and often with a lid, is called a sauté pan or sauté. While a sauté pan can be used like a frying pan, it is designed for lower heat cooking methods, namely sautéing.

Traditionally, frying pans were made of cast iron. Although cast iron is still popular today, especially for outdoor cooking, most frying pans are now made from such metals as aluminium or stainless steel. The materials and construction methods used in modern frying pans vary greatly and some typical materials include: Aluminium/anodized aluminium, cast iron, copper, stainless steel, clad stainless steel with an aluminium or copper core.

A coating is sometimes applied to the surface of the pan to make it non-stick. Frying pans made from bare cast iron or carbon steel can also gain non-stick properties through seasoning and using.

Non-stick pans

A process for bonding teflon to chemically-roughened aluminum was patented in France by Marc Gregoire in 1954. In 1956 he formed a company to market non-stick cookware under the "Tefal" brand name. The durability of the early coatings was poor, but improvements in manufacturing have made these products a kitchen standard. The surface is not as tough as metal and the use of metal utensils (e. g. spatulas) can permanently mar the coating and degrade its non-stick property.

For some cooking preparations a non-stick frying pan is inappropriate, especially for deglazing, where the residue of browning is to be incorporated in a later step such as a pan sauce. Since little or no residue can stick to the surface, the sauce will fail for lack of its primary flavoring agent.

Non-stick frying pans featuring teflon coatings must *never* be heated above 240℃ (464℉), a temperature that can easily be reached in minutes. At higher temperatures non-stick coatings decompose and give off toxic fumes.

Electric frying pans

An electric frying pan or electric skillet incorporates an electric heating element into the frying pan itself and so can function independently off a cooking stove. Accordingly, it has heat-insulated legs for standing on a countertop. (The legs usually attach to handles.) Electric frying pans are common in shapes that are unusual for "unpowered" frying pans, notably square and rectangular. Most are designed with straighter sides than their stovetop cousins and include a lid. In this way they are a cross between a frying pan and a sauté pan.

A modern electric skillet has an additional advantage over the stovetop version: heat regulation. The detachable power cord/unit incorporates a thermostatic control for maintaining the desired temperature.

With the perfection of the thermostatic control, the electric skillet became a popular kitchen appliance. Although it largely has been supplanted by the microwave oven, it is still in use in many kitchens.

Unit 5 Other Cooking Utensils

Learning goals

To master names of other cooking utensils in the kitchen.

To know the usage of cooking utensils.

To apply these names of cooking utensils in expressions and dialogues.

Vocabulary

whisk [wɪsk] *n.* & *v.* 打蛋器;搅拌

grater ['greɪtə(r)] *n.* 擦子

grinder ['graɪndə(r)] *n.* 磨工;研磨器

bone saw 骨锯

steel [stiːl] *n.* & *v.* & *adj.* 钢;钢铁;使……硬如钢;钢的;坚强的

hot pad 隔热垫

hot glove 隔热手套

tenderizer ['tendəraɪzə] *n.* 嫩化剂;嫩肉机

slicer ['slaɪsə] *n.* 切片机

blender ['blendə(r)] *n.* 混合器;搅拌器;果汁机

casserole ['kæsərəʊl] *n.* 砂锅

egg boiler 煮蛋器

toaster ['təʊstə(r)] *n.* 烤面包机

microwave ['maɪkrəweɪv] *n.* & *vt.* 微波;微波炉;用微波炉加热

dish washer 洗碗机

steamer ['stiːmə(r)] *n.* 蒸笼

Dialogue

(*C1=Commis*, *C2=Chef*.)

C1: What is the machine?

C2: It is an automatic dish washer.

C1: How does it work?

C2: First, the dishes are sprayed. Then they are washed with special detergents. They are washed at 140 degrees. They are rinsed at 180 degrees.

C1: Why so hot?

C2: So the dishes will dry by themselves.

C1: Oh, that's it.

Activity 1

Task 1 Try to write names in English below the pictures.

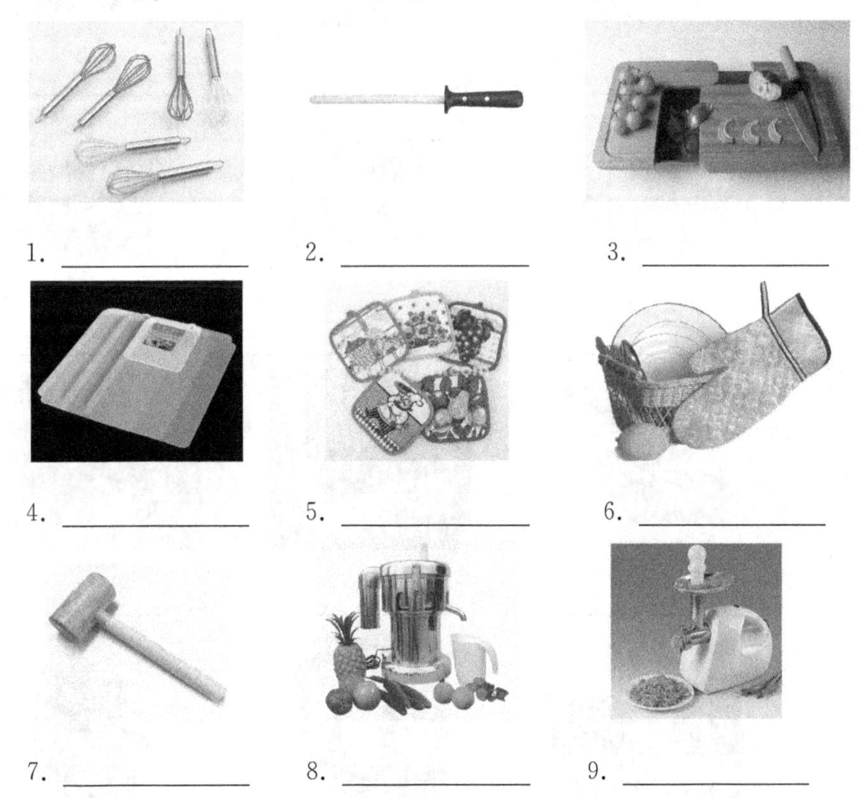

1. ____________ 2. ____________ 3. ____________

4. ____________ 5. ____________ 6. ____________

7. ____________ 8. ____________ 9. ____________

10. ________ 11. ________ 12. ________

13. ________ 14. ________ 15. ________

16. ________ 17. ________ 18. ________

19. ________ 20. ________ 21. ________

22. ________ 23. ________ 28. ________

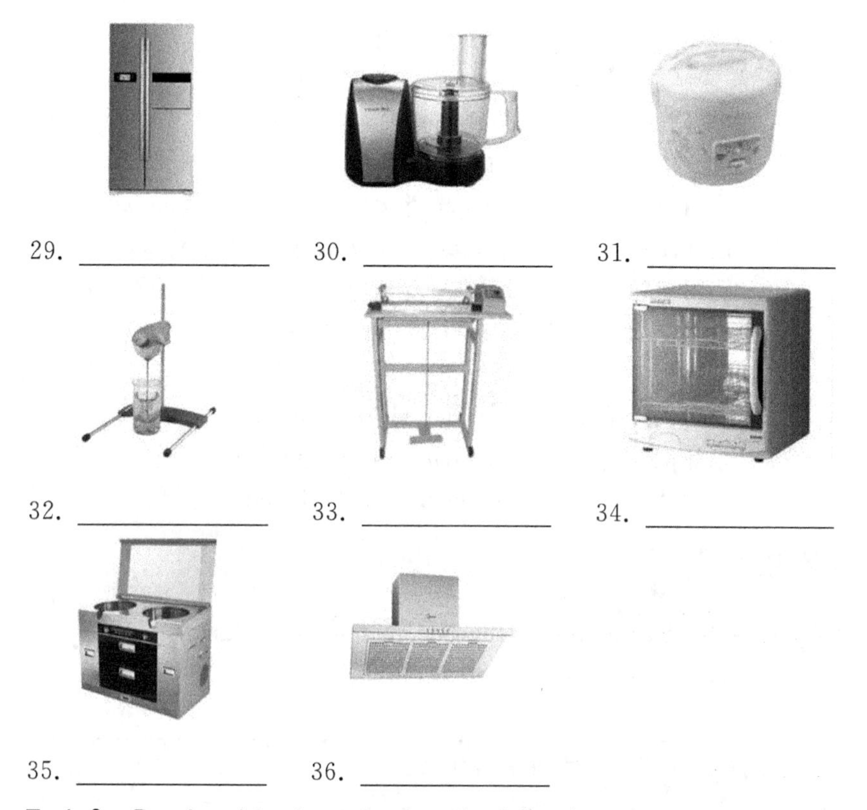

Task 2 Read and try to remember the following expressions about the kitchenware.

1. 使用打蛋器	Use an eggbeater/a whisk. Beat the ingredients with an eggbeater/a whisk.
2. 询问研磨器	Where is the grater?
3. 告知要磨的东西	We need some grated cheese. I will grate the cheese right now.
4. 将骨头对半切开	This is a big bone. I want to cut it in two.
5. 询问其他器具	What should I use then?
6. 告知使用切骨锯	Use a bone saw. We never use knives to cut big bones. It's dangerous. We always use a bone saw. An electric bone saw is safer.

7. 使用磨刀棒	Use a steel to sharpen it.
8. 需要砧板	I need a chopping board to chop on.
9. 木制砧板或塑胶砧板	Do you want a chopping board made of wood or plastic?
10. 使用防热帖	Use a hot pad/glove to take a pan out of the oven.
11. 正确使用防热帖	I will always use dry oven gloves or dry hot pads.
12. 使用肉锤	Should I beat it with this mallet?
13. 使用自动嫩肉机	Use the automatic tenderizer.
14. 切片机	It's a slicer.
15. 绞肉机	You can call it grinder.
16. 果汁机	Put them in the liquidizer. Put them into the blender.

Activity 2

Task 1 Translate the following sentences into English.

1. 请把橙子放入果汁机。

____

2. 请问你用电切片机来做什么?

____

3. 用餐盘罩盖上。

____

4. 我会用海绵和肥皂水把它清洗干净。

____

5. 用洗碗机洗碗能洗干净吗?

____

6. 用隔热手套把烤箱里的平底锅拿出来。

____

Task 2 Look at the pictures given in Task 1, Activity 1. Work in pairs and find out the proper things when you want to make the following dishes.

Cakes: ______________________________

Toasts: ______________________________

Egg wash: ______________________________

Chicken soup: ______________________________

Steamed buns: ______________________________

Meat paste: ______________________________

Apple juice: ______________________________

Activity 3

Task Reading comprehension.

Directions: Work in pairs, read the following passage and try to find out the microwave oven's features.

Microwave Oven

A microwave oven, often colloquially shortened to microwave, is a kitchen appliance that heats food by bombarding it with electromagnetic radiation in the microwave spectrum causing polarized molecules in the food to rotate and build up thermal energy in a process known as dielectric heating. Microwave ovens heat foods quickly and efficiently because excitation is fairly uniform in the outer 25—38 mm of a dense (high water content) food item; food is more evenly heated throughout (except in thick, dense objects) than generally occurs in other cooking techniques.

Percy Spencer invented the first microwave oven after World War II from radar technology developed during the war. Named the "Radarange", it was first sold in 1947. Raytheon later licensed its patents for a home-use microwave oven that was first introduced by Tappan in 1955, but these units were still too large and expensive for general home use. The countertop microwave oven was first introduced in 1967 by the Amana Corporation, which was acquired in 1965 by Raytheon.

Microwave ovens are popular for reheating previously cooked foods and cooking vegetables. They are also useful for rapid heating of otherwise slowly prepared cooking items, such as hot butter, fats, and chocolate. Unlike conventional ovens, microwave ovens usually do not directly brown or caramelize food, since they rarely attain the necessary temperatures to produce Maillard reactions. Exceptions occur in rare cases where the oven is used to heat frying-oil and other very oily items (such as bacon), which attain far higher temperatures than that of boiling water. The boiling-range temperatures produced in high-water-content foods give microwave ovens a limited role in professional cooking, since it usually makes them unsuitable for achievement of culinary effects where the flavors produced by the higher temperatures of frying, browning, or baking are needed. However, additional kinds of heat sources can be added to microwave packaging, or into combination microwave ovens, to produce these other heating effects, and microwave heating may cut the overall time needed to prepare such dishes. Some modern microwave ovens may be part of an over-the-range unit with built-in extractor hoods.

Chapter 3

Conversations in the Kitchen

Unit 1 Expressions on Asking and Answering Between the Chef and Commis

Learning goals

To use the expressions on asking and answering between chefs and commis.

To apply the expressions into conversations.

Vocabulary

cornstarch [ˈkɔːnstɑːtʃ] *n.* 玉蜀黍淀粉

strip [strɪp] *v.* 剥去;剥夺;

vinaigrette [ˌvɪnɪˈgret] *n.* 香料饰盒;色拉调味汁(用油、醋和各种香草等混合而成)

omelet [ˈɒmlɪt] *n.* 煎蛋;鸡蛋卷

tangerine [ˌtændʒəˈriːn] *n.* 橘子;橘子树

preheat [ˌpriːˈhiːt] *v.* 预先加热

approximately [əˈprɒksɪmətli] *adv.* 大约;近似地

margarine [ˈmɑːdʒərɪn] *n.* 人造黄油

skillet [ˈskɪlɪt] *n.* 煎锅

constantly [ˈkɒnstəntli] *adv.* 不断地;经常地

occasionally [əˈkeɪʒnəli] *adv.* 偶尔地

stuff [stʌf] *vt*. 塞满;填满
artichoke [ˈɑːtɪtʃəuk] *n*. 洋蓟;朝鲜蓟;菊芋
mackerel [ˈmækrəl] *n*. 鲭;鲭鱼
soggy [ˈsɒgɪ] *adj*. 湿而软的;潮湿的;湿透的;乏味的
bouillon [ˈbuːjɒn] *n*. 牛肉汤;肉汤
skip [skɪp] *v*. 跳过;略过;遗漏
substitute [ˈsʌbstɪtjuːt] *v*. & *n*. 代用品;代替者;用……代替
salamander [ˈsæləmændə(r)] *n*. 蝾螈;火蜥蜴;能耐高温的物件
overcook [ˈəuvəˈkuk] *vt*. 煮过度

Expressions

1. 询问要做什么	What are we going to do?
	What should I do?
	What shall I do?
	What do you want me to do?
	Do you want me to do something?
	What now?
	What exactly should I do?
2. 做沙拉	We'll make a salad.
	I want you to make a salad.
	We are going to make a salad.
3. 做酱料	Make some eggplant dip, please.
	We're making garlic butter today.
4. 做汤	We are going to make soup with the potatoes.
5. 动作(切)	Slice the mushrooms, please.
6. 动作(去皮)	Peel the onions and garlic.
7. 动作(打蛋)	Break three eggs, please. Mix the eggs with salt and pepper.
8. 动作(混合)	Please mix thirty tablespoons of cornstarch with thirty tablespoons of milk.
	Mix ten cups of milk with five cups of whipping cream.

9. 动作(搅拌)	Stir the first mixture into the second mixture.
	Pour the cream sauce slowly. Keep beating the mixture.
10. 询问是否先做某物	Should I start with the beef?
11. 先做某事	Yes. Please cut the beef into strips.
	Start with the vinaigrette dressing, please.
12. 询问要做哪种蛋饼	What kind of omelet should I fix/cook?
13. 告知哪类蛋饼	A cheese omelet.
14. 告知客人点哪种蛋饼	They ordered a ham and cheese omelet.
15. 询问是否要做某菜	Are we going to make a stew?
16. 确认答案	Yes.
17. 询问先做什么	What should I do first?
	What should we start with?
18. 要拿茄子	Go and get fifteen eggplants.
19. 要切洋葱	Cut the onions into fourths.
20. 询问做某事了没	Have you broken three eggs?
	Have you made the pastry?
	Have you lighted the oven?
21. 表示还没	No. Not yet.
22. 吩咐把还没做好的事完成	Light the oven, please. Cook the fish at four hundred degrees.
23. 已做好	I have already had the eggplants.
	I have already peeled the garlic.
	I have already done it.
	Yes. I have put the pastry in the pie tins.
	OK. I'm finished.
24. 告知接下来要做什么	Good. Then prick the eggplants with a fork.
	Good. Next, peel ten cloves of garlic.
25. 告知也完成了	I have done that, too.
26. 告知还要准备柠檬汁	Excellent! We also need lemon juice.
27. 告知还要使用搅拌机	Put the eggplants, onions, garlic, lemon juice, and olive oil in the blender

	(liquidizer).
28. 告知还要准备馅料	Prepare the cherry filling, please.
29. 询问需要物品	What do you need?
	What will you use?
	What should I hand you?
30. 拿原料	Bring me some tangerines, please.
	Give me three kilos of cherries, please.
	Please get me three cans of imported French mushrooms.
31. 拿用具	I need a flour sieve.
	I will use a flour sieve.
	Hand me a flour sieve, please.
32. 询问要加什么	What should I add now?
33. 告知要加什么	Add the butter, water and rice.
	Add the bouillon.
34. 询问加多少分量	How much lemon juice?
35. 各加多少	How much of each?
36. 要求参照食谱	Look at the recipe, please.
37. 询问食谱在哪里	Where is the recipe?
38. 告知食谱位置	The cookbook is over there.
39. 询问要加几汤匙	How many tablespoons of olive oil should I add?
40. 告知匙数	Twenty tablespoons.
	Just a dash.
41. 询问要加几杯	How many cups of lemon juice should I add?
42. 告知杯数	Two and a half cups of lemon juice.
	Four and a quarter cups of lemon juice.
43. 询问是否要加盐	And should I add some salt?
44. 告知加盐量	Yes. Add 1 teaspoon of salt.
45. 提醒别加太多	Go easy on the salt.
46. 盖上搅拌器盖子	I'll cover the blender.
47. 告知搅拌器用的速度	Blend at high speed.

	Blend at medium speed.
	Blend at low speed.
48. 预热烤箱	I will preheat the oven to three hundred and fifty degrees.
49. 告知将食物放入烤箱	Put the eggplants in the oven, please.
50. 询问温度	At what temperature?
51. 告知温度	At four hundred degrees.
52. 调低温度	Please reduce the temperature to three hundred degrees.
53. 询问时间	For how long?
	How long?
	For how many seconds?
	For how many minutes?
	For how many hours?
54. 告知时间	For fifteen seconds.
	For fifty-five minutes.
	For two hours.
	Thirty minutes approximately.
55. 时间到了	The fifteen minutes are up.
56. 询问火的大小	Over low heat?
	Will you use a low flame?
57. 告知火的大小	No. Over medium heat.
	No. Over low heat.
58. 询问要用什么煮	In what should I cook this chicken?
	How will we cook them?
59. 用油煮	Cook it in butter.
	Cook it in margarine.
	Cook it in vegetable oil.
	Cook it in fat.
60. 用西红柿酱汁煮	We'll cook them in tomato sauce.
61. 询问是否用水煮	In water?
62. 用罐头内的水煮	Yes. Use the water from the can.

63. 遵守用罐头内的水煮的指导	I'll use the water from the asparagus can.
64. 询问是否用锅具煮	Should I cook them in a skillet (frying pan)?
65. 用锅具煮	Yes. Cover and simmer the mushrooms, onions, and garlic for eight minutes.
66. 将锅具盖子盖上	I will cover the beef that is in the skillet.
67. 将锅里菜肴舀起	Remove the vegetables from the skillet.
68. 搅拌频率	Constantly stir it. Occasionally stir it. Stir it once or twice. Stir it three times.
69. 询问搅拌时间	For how long should I stir it? Ten to twelve minutes.
70. 询问是否要煮沸	Should I heat the mixture to a boil?
71. 要煮到滚	I will heat the soup to a boil.
72. 餐点煮沸了	The cream sauce is boiling now.
73. 持续煮沸	Let it boil for one minute.
74. 转小火慢炖	Fine. Then reduce the heat. Cover the soup and simmer it. Bring the water to a boil. Then lower the flame.
75. 提醒别煮过头	Don't let the ingredients get over boiled.
76. 要自然冷却	Let the cream sauce cool slightly. Put the cakes on wire racks to cool.
77. 已冷却	The cakes have cooled.
78. 询问奶油熔化没	Is the butter browning/melting?
79. 奶油转褐色了	The butter is browning.
80. 奶油转淡褐色了	The butter is light brown.
81. 奶油转中褐色了	The butter is medium brown.
82. 奶油转深褐色了	The butter is dark brown.
83. 吩咐快速放入菜肴并搅拌	Pour in the eggs…and stir quickly!
84. 询问是否使用某器具	Should I use a wooden spoon?

85. 表示用其他器具	No. Use a fork.
86. 吩咐冷藏	Refrigerate the Italian dressing.
87. 询问冰多久	For how long should I refrigerate it?
88. 告知时间	Until it is time to serve the salad.
89. 询问接下来步骤	What should I do next?
	And then?
	And then what?
	What next?
	And after that?
	Anything else?
90. 询问同一时间做什么事	And what should I do in the meantime?
	What should I do while the cream sauce is cooling?
91. 询问是否立即行动	Right now?
92. 告知快一点	Yes, hurry up.
	Yes, in a minute.
93. 询问他人在做什么	What are you doing?
94. 告知在做什么事	I'am beating the meat with a cutlet bat.
95. 是否该削皮	Should I peel the potatoes?
96. 是否该浸泡	Should I soak the cauliflower in salted water?
97. 是否该水煮	Should I boil the cauliflower?
	Should I boil it in water with lemon juice?
98. 是否该剥壳	Should I shell the peas from the pea pods?
99. 是否该使用锅具	Should I use a colander?
100. 是否使用某原料	Should I use the peas for a stew?
101. 是否将菜色搭配在一起	Should I serve the peas with butter?
102. 告知需要	Yes, please.
	Yes. Sure.
	Yes. Go right ahead.
103. 不必要	No. It's not necessary.
	No. Not now.
	No. Please wait a minute.

	No. Not today.
	No. Don't bother.
104. 询问先准备什么材料	What should we start with?
105. 告知材料	The spring onions.
106. 询问使用原因	Why?
	Why do you want the eggplants?
107. 询问食材用来做什么	What are we going to do with the big peas?
	And the little peas?
	What are they for?
108. 使用某原料原因	We need finely chopped onions for the soup.
	We'll make a stew.
	Use them for stews.
	The mushrooms are for the stew.
	To serve them with butter.
109. 询问做法原因	Why are you doing that?
110. 某料理动作原因	To peel them easily.
111. 询问是否为罐头、冷冻或新鲜蔬菜	Are the peas frozen or canned (tinned)?
	Are the carrots frozen, canned (tinned), or fresh?
112. 新鲜的	They're fresh.
	Neither. We have fresh peas today.
113. 罐头的	They're canned.
114. 冷冻的	They're frozen.
115. 询问原料用处	What will we use the tomatoes for?
116. 告知原料用处	Tomato soup.
	For the salad.
117. 询问所需原料	Do we need onions for the potato salad?
118. 询问某餐点名称意义	What are crepes?
119. 名称意义	They are thin French pancakes.
120. 询问如何煮	How should I boil them?
	How will we cook them?
	How should I cook them?

121. 询问如何切	How should I cut them open?
	How should I cut it?
	How should I cut up the meat?
	How should I slice them?
122. 询问如何去籽	How should I remove the seeds?
123. 询问如何塞入	How should I stuff them?
124. 询问如何炖	How do you braise beef?
125. 询问如何搅拌	How should I mix the ingredients?
	How should I stir the ingredients?
126. 询问如何上菜	How should I serve the potatoes?
127. 询问如何清理	How should I clean the counter?
128. 询问如何磨刀	How should I sharpen my knife?
129. 询问如何拿起锅子	How should I lift the braising pan?
130. 询问如何处理蔬菜	What should I do with the carrots?
	What about the artichoke leaves?
	How should I cook these potato chips?
131. 询问如何处理鱼类	What should I do with this fish?
	What should I do with the mackerel?
	What should I do with the salmon?
	What should I do with the carp?
132. 询问如何处理奶蛋类	What should I do with cheddar?
	What should I do with the eggs?
133. 表示疑问	Sorry. I didn't understand.
134. 请人示范	Can you show me how to use it?
	Could you explain again how to make the stuffing?
135. 大厨示范	I will demonstrate it.
	I will show you.
	Like this.
	Watch me.
	Cut them into small pieces.
136. 请求有机会试试看	Let me try them out.

137. 同意一试	No problem.
138. 成品失败,再做一次	This one failed. Can we make another one?
139. 询问步骤哪里出错	The texture is not right. Where did I go wrong?
140. 询问为什么成果不一样	I did exactly what you taught me. How come mine tastes differently from yours?
141. 告知过量出错	It's soggy because you put in too much water.
142. 告知步骤出错	You won't get the same result if you skip this step. You won't get the same result if you don't follow the steps.
143. 询问是否听懂	Are you with me? Do you follow me? Do you get it? Do you understand? See what I mean?
144. 表示感到很生疏	I have never done this before.
145. 告知很简单	It's easy.
146. 告知能以别种材料代替	You can replace/substitute the meat with tofu if you want to prepare a vegetarian dish.
147. 询问要切什么	What should I cut up?
148. 询问要不要切	Should I cut up the potatoes?
149. 询问要不要清洗	Should I wash them?
150. 解释清洗原因	Cauliflower is always dirty.
151. 同意	Yes, go ahead.
152. 不同意	No, please don't.
153. 询问是否要加热	Should I heat the asparagus?
154. 询问为了哪道菜	For the salad?
155. 告知为了那道菜	Yes, they're for the salad.
156. 询问用具	With my cook's knife?

157. 答案正确	That's right.
158. 放手去做	Yes. Go ahead.
159. 是否现在做	Now?
160. 现在就做	Yes. Now, please.
161. 询问接下来要做什么	What are you going to do?
	What are you going to chop?
162. 告知要做的事	I am going to strain peas.
	I am going to serve the sauce.
	I am going to make a sauce.
	I am going to butter vegetables.
	I am going to pick up a hot pan.
	I am going to stir some ingredients.
	I will fry some onions.
	I will transfer some vegetables to a plate.
	I will fry a carton of eggs.
	I will grate some cheese.
	I will sharpen my knife.
	I will ice a cake.
	I will skim some oil.
	I am going to marinate the chicken all right.
	I will leave it in the refrigerator.
163. 询问熟了没有	Are they cooked?
	Is this meat cooked?
	Is the meat done?
	Have you finished cooking the meat?
164. 已经熟了	Yes.
	Yes. The meat is done/cooked.
165. 不知道熟了没	I don't know.
166. 还没熟	It's not done yet.
167. 测温度	Test the meat, please.
168. 询问测量方式	How?

169. 告知测量方式	Stick a meat thermometer in the meat.
170. 测量结果	The meat should be done in about twenty minutes.
171. 告知已煮好	The chicken is ready. The rice is done/cooked/ready.
172. 询问放哪里	Where should I put the chicken?
173. 告知装盘	Let's put it (them) in a soup tureen. Put the bowls in the salamander just before you serve the soup.
174. 一起上菜	We'll serve the leaves with sauce. Remember to serve the carrots with butter.
175. 淋酱时机	Pour the dressing on the salad just before serving.
176. 装饰盘面	I will garnish the beef stroganoff with some parsley.
177. 提醒塞上水槽塞子	Before you wash the green peppers in the sink, stop up the drain with the stopper.
178. 表示明白	I see. OK. Right.
179. 命令做某事	Wash the tomatoes!
180. 告知蛋糕状态	The cake is ready. The cake is firm. The cake is done. The cake is baked.
181. 告知口味	It's sweet. It's heavy. It's light.
182. 告知口感	It's tender. It's tough. It's raw. It's cold.

	It's hot.
183. 询问蔬菜变色原因	What makes vegetables lose their colors?
184. 告知蔬菜变色原因	Overcooking them.
185. 询问维生素流失原因	What makes vegetables lose their vitamins?
186. 告知维生素流失原因	Cutting them into small pieces.
187. 询问蔬菜最佳的烹调方式	What is the best way to cook vegetables?
188. 告知方式	In small quantities, in a steam kettle or pressure cooker.
189. 询问英文说法	What do you call this in English? What is the English word for this? What do you call "mint" in English? What is "mint" in English? What do you call this dish in English?
190. 上菜注意事项	Always serve equal portions.
191. 节省原料	Reuse butter, cream, bread, etc.
192. 时常检查冰箱	Check the refrigerator often.
193. 不浪费热水	Do not waste hot water.
194. 先使用放了较久的食物	Use old food first.

Unit 2 Conversation Examples in the Kitchen

Learning goals

To apply what has been learned into kitchen conversations.

To make easy conversations with others in the kitchen.

Vocabulary

awesome [ˈɔːsəm] *adj.* 可怕的；表示敬畏的

renovate [ˈrenəveɪt] *vt.* 修理；翻新；革新

apple crumble *n.* 奶油甜脆苹果饼

mess [mes] *n.* & *v.* 混乱;混杂;弄糟;弄乱
alcohol [ˈælkəhɒl] *n.* 酒精;酒;乙醇
evaporate [ɪˈvæpəreɪt] *v.* 蒸发;失去水分;消失
breadcrumbs [ˈbredkrʌmz] *n.* & *v.* 面包屑;在……上覆以面包屑
worktop [ˈwɜːktɒp] *n.* (厨房柜橱上的)操作面
leftover [ˈleftəʊvə(r)] *n.* & *adj.* 剩饭;残留物;残余的
rag [ræg] *n.* 破布;碎布
cupboard [ˈkʌbəd] *n.* 碗橱;衣柜
appliance [əˈplaɪəns] *n.* 器具;应用;装置;器械
spaghetti [spəˈgeti] *n.* 意大利式细面条
linguine [ˈlɪŋgwiːniː] *n.* (意大利)扁面条
marble [ˈmɑːbl] *n.* & *v.* 大理石;弹子;使有大理石的花纹
crust [krʌst] *n.* 外壳;面包皮
topping [ˈtɒpɪŋ] *n.* 糕点上的装饰配料
self-rising *adj.* 渗进发酵物质的
dissolve [dɪˈzɒlv] *v.* 溶解;液化
elastic [ɪˈlæstɪk] *n.* & *adj.* 橡皮圈;灵活的;有弹性的

Dialogue

Example 1

A: Wow! Your kitchen is awesome.
B: Thank you. It's newly renovated.
A: Oh, you've bought a new coffee maker.
B: Yes. Let's brew some coffee with it.
A: Great. How long shall we wait?
B: About 15 minutes. Michel, do you like red bean soup?
A: Yes, it's so cool.
B: We can cook it now.
A: It'll save time if we use the pressure cooker.
B: Oh, what a pity! My pressure cooker doesn't work now. But I can borrow one from my neighbor.
A: OK.

Example 2

(*Archie is preparing a dessert. Daisy is trying to help. D=Daisy, A=Archie.*)

D: So, what are you making for dessert?

A: Apple crumble.

D: What's that?

A: It's a typical English dessert. It's delicious.

D: So, how do you make it?

A: Well, first, you get some apples. Then, you get a peeler and peel them, and then cut them up into fairly small slices. And then you put them into a baking tray.

D: Shouldn't you be wearing an apron?

A: Well, I guess I could, but I'm not planning to make a mess! Anyway, then, you get some flour, sugar and butter. You put it all into a mixing bowl and mix it up with your fingers. This is what's called the crumble.

D: Shouldn't you measure it out properly?

A: Well, I could use the scales, but you don't really need to.

D: Shouldn't you be using white sugar?

A: You can, but I prefer it with brown sugar. Anyway, then I grate in a bit of lemon zest with the grater, plus a bit of orange zest and a bit of nutmeg and cinnamon. I also add a bit of brandy. Hey, can you pass me the bottle opener, please?

D: (She gives him the bottle opener.) Here you are. You shouldn't be suing alcohol, should you? The dessert's for your kids, isn't it?

A: It's OK. It evaporates, more or less. Anyway, once you've got all that in the baking tray, you put the crumble on top like this.

D: Nice!

A: But if you want, you could make it into an apple pie. So, instead of putting the crumble on top, you just put a bit of pastry over the apple mixture, cut off the edges with a spatula and use a brush to coat the pastry with a beaten egg.

D：OK. Thanks. Hey，shall I help you put it in the oven?

A：OK. But be careful because…(Daisy drops the plate.)

D：Whoops! Sorry.

A：…It's quite heavy!

Example 3

(C=Commis cook，V= Vegetable cook.)

C：What shall I do (with the ingredients of the dish)?

V：Wash the cucumbers，tomatoes，onions and lettuce well.

C：OK. The cucumbers are always dirty.

V：Soak them in salt water.

C：For how long?

V：For about 20 minutes.

C：And what shall I do then?

V：Cut the cucumbers into pieces，and dice the tomatoes and onions.

C：What shall I cook the cucumbers for?

V：For vegetable salad.

C：OK.

Example 4

(A kitchen dialogue between a chef and a commis for cooking the dish named Baked Fish with Sesame Oil. C=Commis，B=Brian.)

C：What shall I do with the fish?

B：Scale it，cut off the fins and then gut it.

C：What kind of knife can I use to scale the fish?

B：Use the fish scaler to do it. And use these fish scissors to gut it.

C：So these are special scissors.

B：Yes. Use them to cut open the stomach of the fish. Then take out the guts.

C：OK.

B：After you finish it，cook some onion.

C: Already done.

B: Then mix the sesame seeds, water, garlic, salt, lemon juice and red pepper with a spoon.

C: All right. What next?

B: Sprinkle the baking dish with breadcrumbs and parsley.

C: Shall I put the fish in the baking dish?

B: Yes. Pour the sesame seeds and onions over the fish. Light the oven and cook the fish at 400 degrees.

C: For how long?

B: For 20 to 25 minutes. When it is cooked, garnish the fish with parsley and olives.

Example 5

(*C=Chef*, *A=Assistance.*)

C: Can you give me a hand with some things in the kitchen? I don't think I can finish everything in time.

A: OK, what do you want me to do?

C: First of all, I need you to do the drying up. I've almost finished the washing up. I'm going to clean the cooker when I finish.

A: OK. I'll put the plates and cutlery away as I dry them. Where is the tea towel? Oh, here it is.

C: We'll have this finished in no time with both of us working on it.

A: While you're cleaning the cooker, I'll wipe the worktop. That is a great meat, by the way.

C: Actually, it is just some leftovers from yesterday. I made far too much food to eat alone. I am glad you could come over to help me finish it.

A: My pleasure! This tea towel's a little ragged. Do you have another one?

C: Yes. Look in that drawer. I should throw the old one out.

A: Keep it and use it as rag. You can clean your bicycle with it.

C: So, what's new in the kitchen? That refrigerator is new, isn't it?

A: Yes. I needed a large one. Before, I had a separate refrigerator and freezer, but this has both combined into one.

C: That's usual nowadays. You've added a few shelves too.

A: Yes. You know I've been cooking more kinds of food recently and I needed some extra space for spices and ingredients.

C: Did you buy new cupboards too?

A: No, I didn't. I gave them a really good clean, so they just look new. The worktop was in poor condition, so I had a new one added.

C: I see that you have bought several new pots, pans and other utensils.

A: Yes, I have. I need them to help me with these new dishes I'm trying to make. I need a little more practice before I invite guests over.

C: Looking at the spice rack, I'd say you've been learning how to make Asian food.

A: Yes. I've always liked Indian and Thai food, so I've been trying to make dishes from those countries. I'm pretty good at making curries now, but I still need practice at making Thai food.

C: Both kinds of food are becoming popular.

Example 6

(*Today is the first day of the class "Italian Cooking". The cooking instructor Ms Terry is giving an introduction to the class. T=Ms. Terry, S1=Student 1, S2=Student 2.*)

T: Before we start, I'd like to go over some basic rules. You're welcome to use all the equipment in the kitchen. However, the number of some appliances is limited. I'll put you into three pairs. Each pair will be sharing one set of cooking tools and appliances.

S1: Do we always work with the same person?

T: Yes. Once you are paired, you'll stay with the same partner throughout the whole course. You'll be responsible for cleaning up the equipment that is assigned to you.

S2: Can we leave food in the fridge?

T: Yes, you can. Please put your name on the bag so they don't get mixed up. OK, let's get started. Today we'll be cooking pasta with pesto sauce.

S1: Can I use spaghetti instead of linguine?

T: Yes, that's fine. Now, chop up three or four fresh basil leaves, 2 cloves of garlic, and half cup of pine nuts. Put them in the food processor, which you can find in the cupboard next to the oven.

S2: Can I use my own mortar instead?

T: Just make sure it's marble, not wood. In the meantime, we can start cooking the linguine. Here we're using a gas burner. Please be careful when the pasta water is boiling and rising. Now, I'll be walking around the kitchen. Please let me know if you need any help.

Example 7

(*In the second cooking class, Ms. Terry is teaching her students how to make pizza. T=Ms. Terry, S3=Student* 3, *S4=Student* 4.)

T: Today's class is divided into two parts. First, I'll show you how to prepare pizza crust; then you will learn how to make pizza toppings. The ingredients for the dough are yeast, sugar, salt, vegetable oil, and flour.

S3: Is this self-rising flour?

T: Yes, it is. Now get a cup of warm water. Add yeast to the water and stir until it's dissolved. Then add one teaspoon of sugar, one teaspoon of salt, two teaspoons of oil, and two cups of flour.

S3: Should I keep stirring?

T: Yes. Add another half cup of flour and keep stirring. Now, sprinkle some flour on the board. Knead the dough until it feels elastic. While waiting for the dough to rise, let's prepare the toppings. Basically, you can put whatever you like on your crust.

Here I have sliced Italian sausage, ground beef, peppers, onions, tomato paste, and shredded cheese.

S3: What kind of cheese is it?

T: We usually use Mozzarella cheese. Now set the dough on a greased pan, and layer your toppings onto the pizza dough. Put the whole thing in a preheated oven.

S4: How long should we bake it for?

T: For about 15 minutes.

Chapter 4

Food Materials

Unit 1 Vegetables

Learning goals

To know essential information of various kinds of vegetables, including bulb vegetables, root vegetables, fruit vegetables, leaf vegetables, stem vegetables, tuber vegetables and flower vegetables.

Vocabulary

asparagus [əˈspærəɡəs] *n.* 芦笋
chive [tʃaɪv] *n.* 香葱
scallion [ˈskæliən] *n.* 青葱
garlic [ˈɡɑːlɪk] *n.* 大蒜
shallot [ʃəˈlɒt] *n.* 葱
onion [ˈʌnjən] *n.* 洋葱
water chestnut 菱角,荸荠
turnip [ˈtɜːnɪp] *n.* 萝卜
carrot [ˈkærət] *n.* 胡萝卜
radish [ˈrædɪʃ] *n.* 小萝卜
eggplant [ˈeɡplɑːnt] *n.* 茄子

green pepper 青椒
olive [ˈɒlɪv] n. 橄榄
cucumber [ˈkjuːkʌmbə(r)] n. 黄瓜
okra [ˈəukrə] n. 黄秋葵
tomato [təˈmeɪtəu] n. 西红柿
pumpkin [ˈpʌmpkɪn] n. 南瓜
spinach [ˈspɪnɪtʃ] n. 菠菜
lettuce [ˈletɪs] n. 莴苣
cabbage [ˈkæbɪdʒ] n. 甘蓝
celery [ˈseləri] n. 芹菜
potato [pəˈteɪtəʊ] n. 马铃薯
cauliflower [ˈkɒliflaʊə(r)] n. 花椰菜
red beet 甜菜根
winter melon 冬瓜
bitter melon 苦瓜
purslane [ˈpɜːslɪn] n. 马齿苋
arugula [æˈruːgjʊlə] n. 芝麻菜
cress [kres] n. 水芹;水蔚菜
radicchio [ræˈdiːkiəʊ] n. 菊苣
curly endive 皱叶苦苣
escarole [ˈeskərəʊl] n. 茅菜(做沙拉的一种蔬菜)
endive [ˈendaɪv] n. 菊苣
celtuce [ˈkeltʃjuːs] n. 青菜
kale [keɪl] n. 羽衣甘蓝;无头甘蓝
salad savoy 沙拉萨瓦
cardoon [kɑːˈduːn] n. 刺棘蓟;刺菜蓟
rhubarb [ˈruːbɑːb] n. 大黄
fennel [ˈfenl] n. 茴香;小茴香
artichoke [ˈɑːtɪtʃəuk] n. 朝鲜蓟;洋蓟
bean [biːn] n. 豆
pea [piː] n. 豌豆,豌豆类

Dialogue

(*S=Steven*, *G=George.*)

S: George, can you do me a favor?

G: Of course, Steven.

S: Help me to peel all of these tomatoes as soon as possible.

G: All of them?

S: Yes, all of them.

G: But how should I do?

S: Listen, firstly, cut an X into the bottom of the tomato, and be sure not too deep.

G: OK, and then?

S: After that, drop it into boiling water for about 10 to 15 seconds.

G: 10 to 15 seconds, okay.

S: Then, take it out of boiling water and plunge (投入) it into ice water. See, it is easy to use a paring knife to peel it now.

G: I see. I'll do my best.

S: Thank you very much.

G: You are welcome.

Activity 1

Task Try to write the names in English below the pictures.

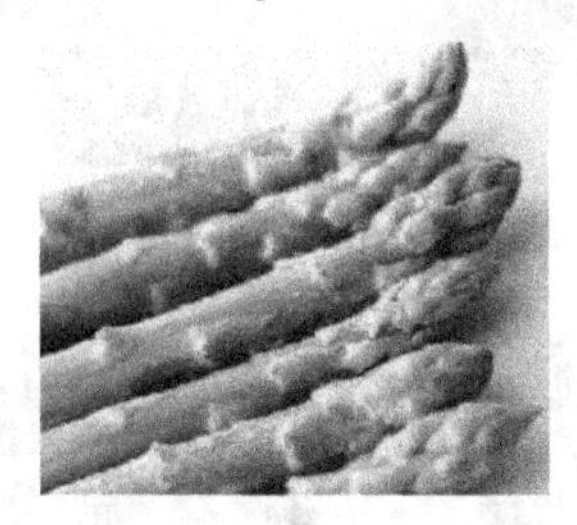

1. 
2. ______________
3. ______________

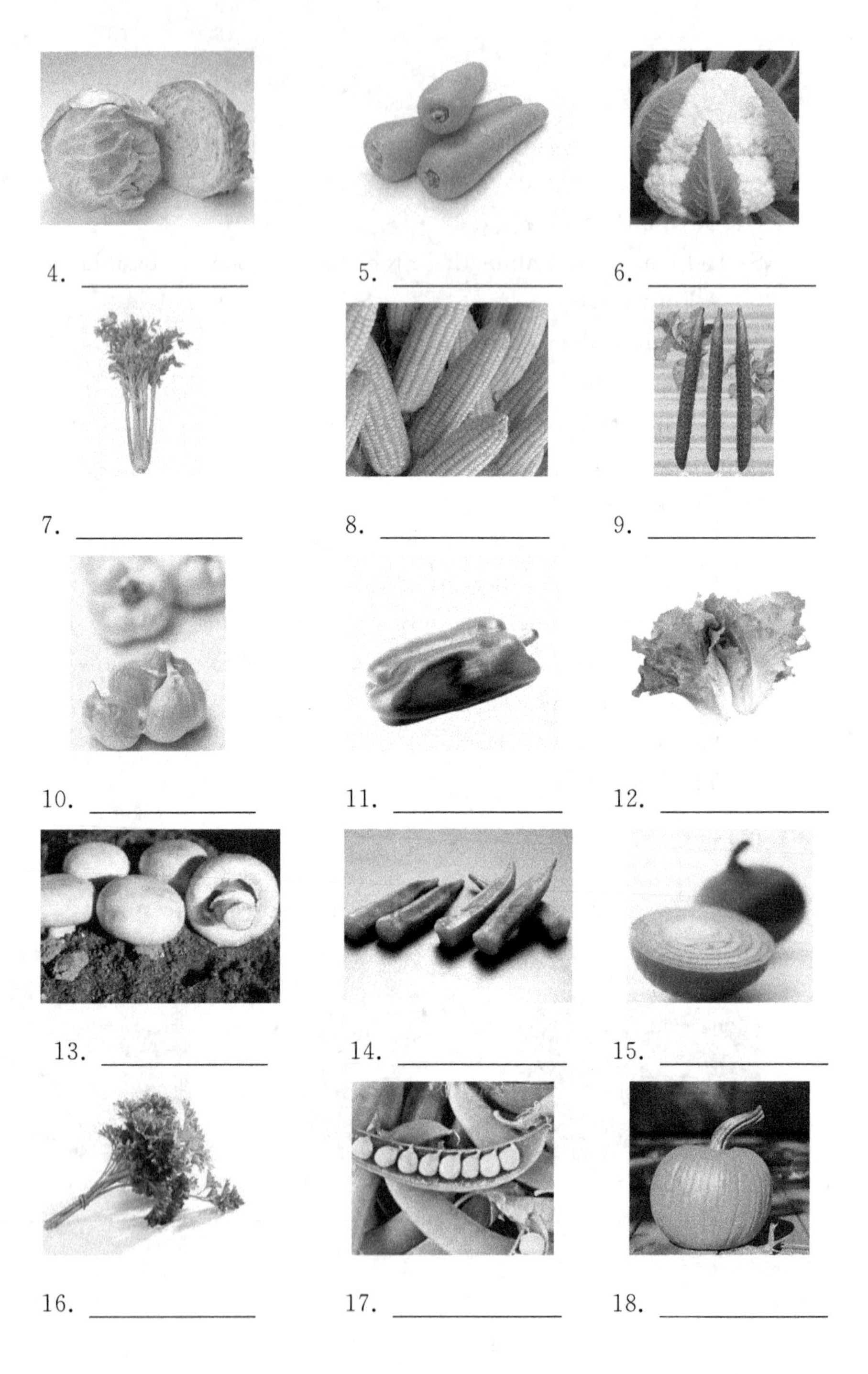

4. ______________ 5. ______________ 6. ______________

7. ______________ 8. ______________ 9. ______________

10. ______________ 11. ______________ 12. ______________

13. ______________ 14. ______________ 15. ______________

16. ______________ 17. ______________ 18. ______________

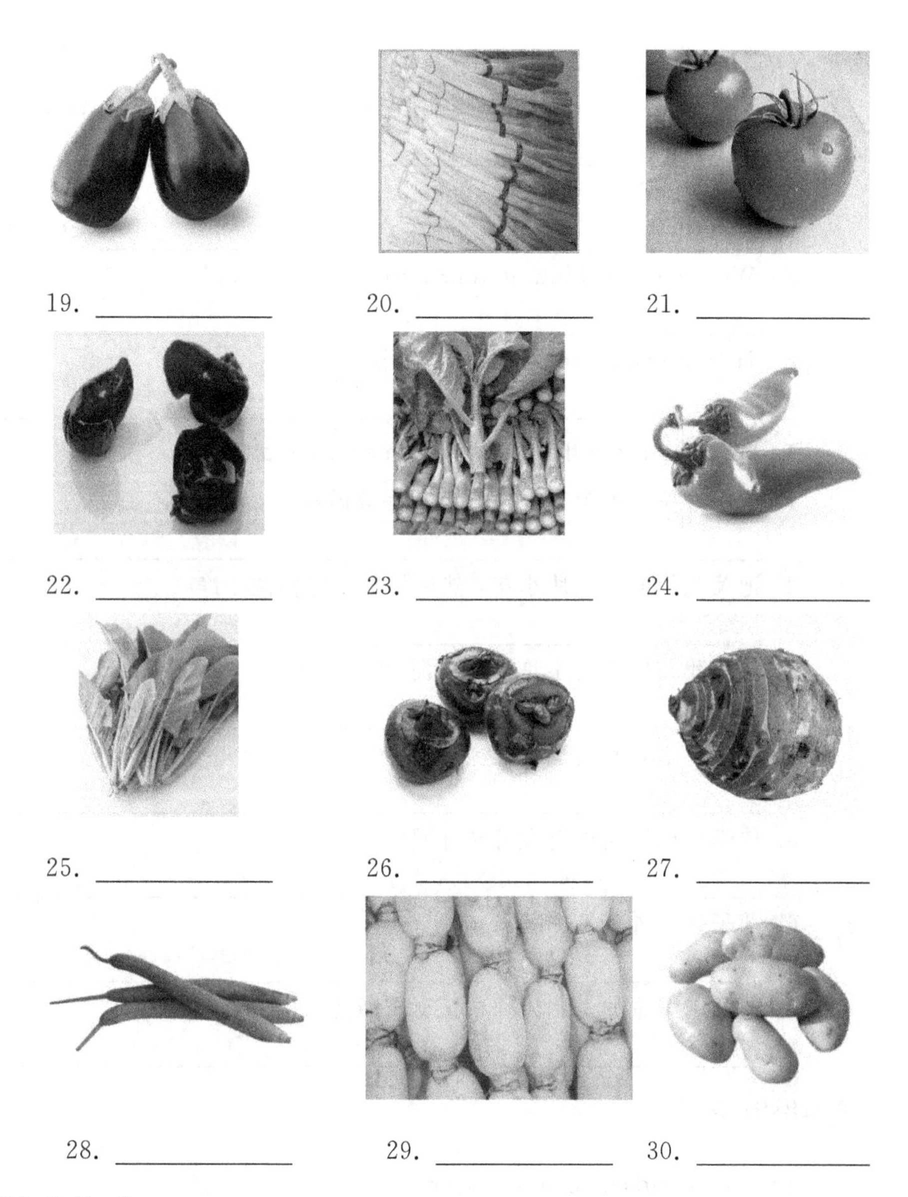

19. ____________ 20. ____________ 21. ____________

22. ____________ 23. ____________ 24. ____________

25. ____________ 26. ____________ 27. ____________

28. ____________ 29. ____________ 30. ____________

Activity 2

Task 1 Suppose you are John and you are a potman. You are interviewed in the restaurant where you work. Answer the reporter's questions and fill in the blanks.

(*R*=*Reperter*, *J*=*John.*)

R: Tell me a bit about your job. What do you do in the restaurant?

J: ____________________

R: I see, and what do you do as a potman?

J: ____________________

R: What do you think of your job?

J: ____________________

R: How do you find a potman's job?

J: ____________________

Task 2 Translate the following sentences into English.

1. 这份沙拉是由苹果、梨、土豆和芹菜做成的。

____

2. 把菜花掰成一块块小花，然后仔细把它们洗干净。

____

3. 请给我花椰菜而不是马铃薯泥，可以吗?

____

4. 将3杯水煮沸后加盐及糖，将芦笋烩熟。

____

5. 你尝得出炖肉里有大蒜味儿吗?

____

6. 茄子是一种紫色的蔬菜。

____

7. 他们用沉重的木制压榨机把橄榄压碎。

____

Activity 3

Task 1 Reading comprehension.

American Cooking Style

Do you think American cooking is only about opening a bag of food and putting it into the microwave oven? If you think so, you are wrong. It is true that many Americans eat bread for breakfast, sandwiches for lunch and fast dinners. Many Americans like them because they can eat them in 10 minutes or less. But many Americans think that cooking skills are necessary. Parents—especially mothers—think it is important to let their children—especially daughters—learn how to cook. Most Americans think there's nothing better than a good meal cooked at home.

Every cook has his or her own cooking skills. But there are some skills that most people use. For example, Americans often bake to make food.

In a big meal, there will be meat, a few vegetables, some bread and often a dessert. Americans also like to make the meal colorful. Having some different colors of food usually makes for a healthy meal.

Question 1: What do you think about American cooking style after reading the passage?

Question 2: What kinds of food should be prepared in a big meal in America?

Task 2 Discuss the following topic with your partner.

Do you know what is a food critic, and what should you have to be a food critic?

Unit 2 Fruits

Learning goals

To know essential information of various kinds of fruits and the taste of fruits.

To know about the usage of some fruits in Western dishes.

Vocabulary

blueberry [ˈbluːbərɪ] *n.* 蓝莓

cherry [ˈtʃeri] *n.* 樱桃

grape [greɪp] *n.* 葡萄

strawberry [ˈstrɔːbəri] *n.* 草莓
plum [plʌm] *n.* 李子
peach [piːtʃ] *n.* 桃子
date [deɪt] *n.* 枣
apricot [ˈeɪprɪkɒt] *n.* 杏仁
apple [ˈæpl] *n.* 苹果
pear [peə(r)] *n.* 梨
grapefruit [ˈgreɪpfruːt] *n.* 葡萄柚
lemon [ˈlemən] *n.* 柠檬
tangerine [ˌtændʒəˈriːn] *n.* 柑橘
lime [laɪm] *n.* 酸橙
banana [bəˈnɑːnə] *n.* 香蕉
durian [ˈdʊəriən] *n.* 榴梿果
longan [ˈlɒŋgən] *n.* 龙眼
lychee [ˈlaɪtʃiː] *n.* 荔枝
papaya [pəˈpaɪə] *n.* 番木瓜果
kiwi [kiːwiː] *n.* 奇异果
mango [ˈmæŋgəʊ] *n.* 芒果
pineapple [ˈpaɪnæpl] *n.* 菠萝
melon [ˈmelən] *n.* 甜瓜
blackberry [ˈblækbəri] *n.* 黑莓
quince [kwɪns] *n.* 温柏,温柏树
pomelo [ˈpɒmələʊ] *n.* 柚子,文旦
orange [ˈɒrɪndʒ] *n.* 橘子,橙子
mandarin [ˈmændərɪn] *n.* 柑橘
carambola [ˌkærəmˈbəʊlə] *n.* 五敛子树;阳桃树
cherimoya [ˌtʃerɪˈmɔɪə] *n.* 番荔枝(产于南美洲),番荔枝的果实
jackfruit [ˈdʒækfruːt] *n.* 木菠萝
pomegranate [ˈpɒmˌgrænɪt, ˈpɒmɪ-] *n.* 石榴;石榴树
fig [fɪg] *n.* 无花果;无花果树
watermelon [ˈwɔːtəmelən] *n.* 西瓜

Dialogue

(*J=Julia, L=Lisa.*)

J: Do you have any plans for dinner tonight?

L: No, I was thinking of putting a frozen pizza in the oven or something. How about you?

J: I was thinking maybe we could make dinner together tonight. What do you think?

L: I'm absolutely useless at cooking!

J: I can teach you how to cook something healthy. Frozen pizzas are so bad for you!

L: I know they aren't good for health, but they are cheap, convenient, and fairly tasty.

J: I've recently seen a recipe for spicy chicken curry in a magazine. Maybe we can try that?

L: Yeah, why not? Do you have all the ingredients?

J: I bought all the ingredients this morning, so let's start!

L: What do we do first?

J: First, you need to wash the vegetables and then chop them into small pieces.

L: OK. Should I heat the wok?

J: Yes. Once it gets hot, put a little oil in it, add the vegetables and stir fry them for a few minutes.

L: What about the chicken?

J: That needs to be cut into thin strips about 3 cm long and then it can be stir fried on its own until it's cooked through.

L: How about the rice?

J: I'll prepare that. Do you prefer white rice or brown rice?

L: White rice, please. None of that healthy brown stuff for me!

Activity 1

Task　Try to write the names in English below the pictures.

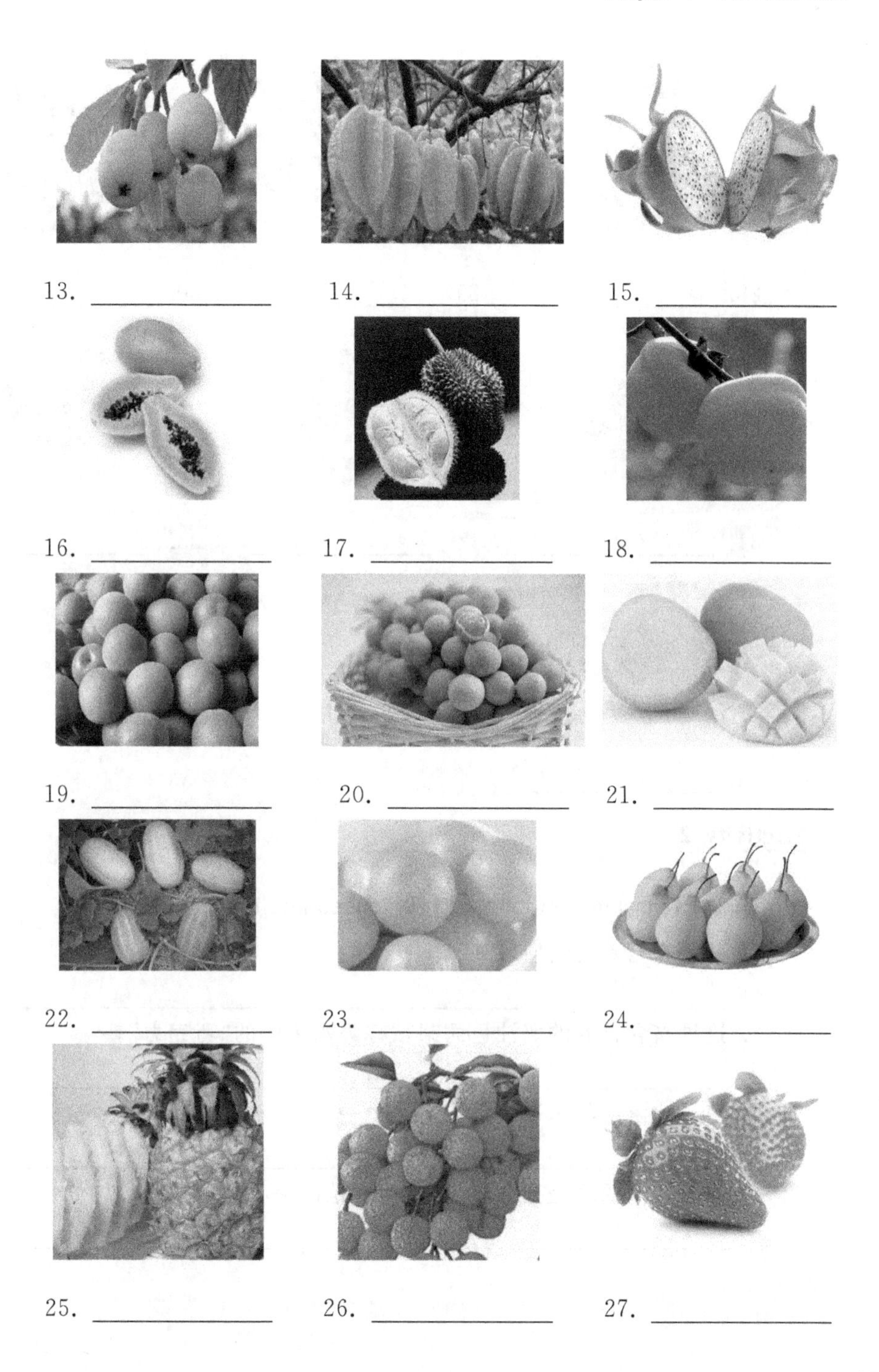
13. ______
14. ______
15. ______
16. ______
17. ______
18. ______
19. ______
20. ______
21. ______
22. ______
23. ______
24. ______
25. ______
26. ______
27. ______

28. ______________ 29. ______________ 30. ______________

31. ______________ 32. ______________ 33. ______________

34. ______________

Activity 2

Task　Translate the following sentences into English.

1. 浆果有营养,并且带甜味。

____

2. 这种富含营养的奶油色鳄梨被称为水果中的“巧克力”。

____

3. 芒果冰淇淋的味道很不错。

____

4. 你所要做的就是把菠萝去皮,然后挖出果心。

____

5. 切一颗奇异果，搭配火鸡肉、木瓜、杏仁片和菠菜叶，就是一道清爽的沙拉了。

____

6. 那盆鱼上配了几片柠檬作为装饰。

____

7. 厨房里好像正在做蓝莓派。

____

8. 他把草莓酱涂在烤面包片上。

____

Activity 3

Task 1 Reading comprehension.

Usage of Fruits

Many hundreds of fruits, including fleshy fruits like the apple, peach, pear, kiwifruit, watermelon and mango, are commercially valuable as human food, eaten both fresh and as jams, marmalade and other preserves. Fruits are also used in manufactured foods like cookies, muffins, yogurt, ice cream, cakes, and many more. Many fruits are used to make beverages, such as fruit juices (orange juice, apple juice, grape juice, etc.) or alcoholic beverages, such as wine or brandy. Apples are often used to make vinegar. Fruits are also used for gift giving: The fruit basket and fruit bouquet are some common forms of fruit gifts.

Many vegetables are botanical fruits, including the tomato, bell pepper, eggplant, okra, squash, pumpkin, green bean, cucumber and zucchini. Olive fruit is pressed for olive oil. Spices like vanilla, paprika, allspice and black pepper are derived from berries.

Question 1: Why do people usually eat fruit?

Question 2: What kinds of food can be manufactured with fruits?

Task 2 Further reading.

Wine

Wine is an alcoholic beverage made from fermentation of grape juice. The natural chemical balance of grapes is such that they can ferment without the addition of sugars, acids, enzymes, or other nutrients. Although fruits other than grapes can also be fermented, the resultant wines are normally named after the fruit from which they are produced (for example, apple wine) and are known as fruit wine (or country wine). Others, such as barley wine and rice wine (e. g. sake), are made from starch-based materials and resemble beer more than wine; ginger wine is fortified with brandy. In these cases, the use of the term "wine" is a reference to the higher alcohol content, rather than the production process. The commercial use of the word "wine" (and its equivalent in other languages) is protected by law in many jurisdictions. Wine is produced by fermenting crushed grapes using various types of yeasts, which consume the sugars found in the grapes and convert them into alcohol. Different varieties of grapes and strains of yeasts are used depending on the types of wine produced.

Wine stems from an extended and rich history dating back to about 8,000 years ago and is thought to have originated in present-day Georgia or Iran. Wine is thought to have appeared in Europe about 6,500 years ago in present-day Bulgaria and Greece and was very common in ancient Greece and Rome; the Greek god Dionysos, and his Roman counterpart Liber represented wine. Wine continues to play a role in religious ceremonies, such as Kiddush in Judaism and the Eucharist in Christianity.

Unit 3 Herbs and Seasonings

Learning goals

To know essential information of various kinds of herbs and seasonings, including the function of herbs and seasonings in dishes.

Vocabulary

dill [dɪl] *n.* 小茴香

anise [ˈænɪs] *n.* 茴芹
chervil [ˈtʃɜːvɪl] *n.* 山萝卜
rosemary [ˈrəuzməri] 迷迭香
oregano [ˌɒrɪˈgɑːnəu] *n.* 牛至
basil [ˈbæzəl, ˈbeɪzəl] *n.* 罗勒属植物
sage [seɪdʒ] *n.* 鼠尾草
lemon thyme 柠檬百里香
clove [kləuv] *n.* 丁香
nutmeg [nʌtmeg] *n.* 肉豆蔻
lemon balm 蜜蜂花
lemon grass 香茅草
saffron [ˈsæfrən] *n.* 藏红花
cumin [ˈkʌmɪn] 孜然芹
curry [ˈkəːri, ˈkʌri] *n.* 咖喱食品
turmeric [ˈtɜːmərɪk] *n.* 姜黄
chili [ˈtʃɪli] *n.* 红辣椒
vanilla pod 香草荚
soy sauce 酱油
white wine vinegar 白葡萄酒醋
sea salt 海盐
table salt 精盐
butter [ˈbʌtə(r)] *n.* 黄油;奶油
mustard [ˈmʌstəd] *n.* 芥末;芥菜
rock sugar 冰糖
chili paste 辣椒酱

Dialogue

(*J*=*Julia*, *L*=*Lisa*.)

L: Do you like cooking, Julia?

J: I really enjoy it, especially when it ends up tasting good!

L: How often do you usually cook?

J: I usually make a few salads for lunch throughout the week and I make dinner about 6 times a week.

L: That's a lot of cooking. You must save a lot of money by eating at home so much.

J: I do. If you cook at home, you can eat healthy food cheaply.

L: What kind of dishes do you usually make?

J: I almost always make either a beef roast or a chicken roast with asparagus, parsnips, peas, carrots and potatoes on Sunday.

L: Do you make a lot of traditional British food?

J: Aside from the Sunday roast, we usually eat bangers and mash, or fish chips once a week.

L: How about spicy food?

J: My family loves spicy food. We often eat Chinese, Thai, Indian, or Mexican food when we're in the mood for spice.

L: What's your favorite dish to make?

J: I absolutely love making mousaka, which is a Greek dish with eggplant. But it takes a lot of time, so I don't often make it.

Activity 1

Task　Try to write the names in English below the pictures.

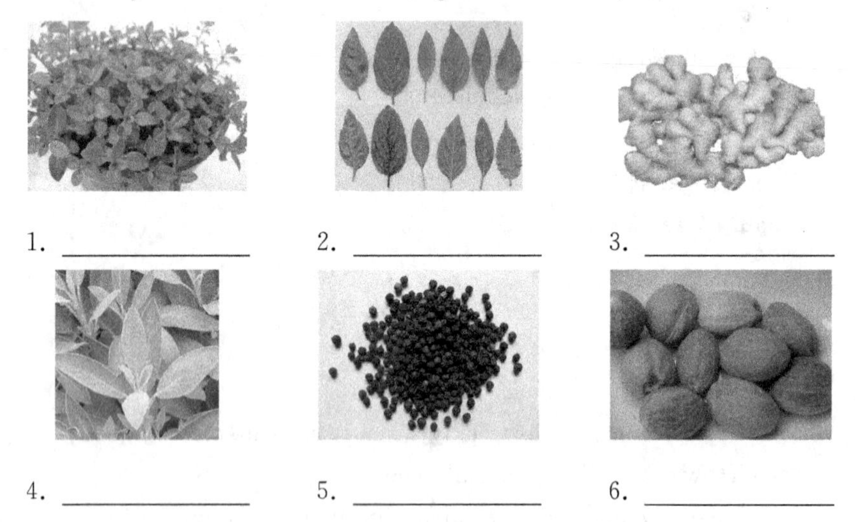

1. ________　2. ________　3. ________

4. ________　5. ________　6. ________

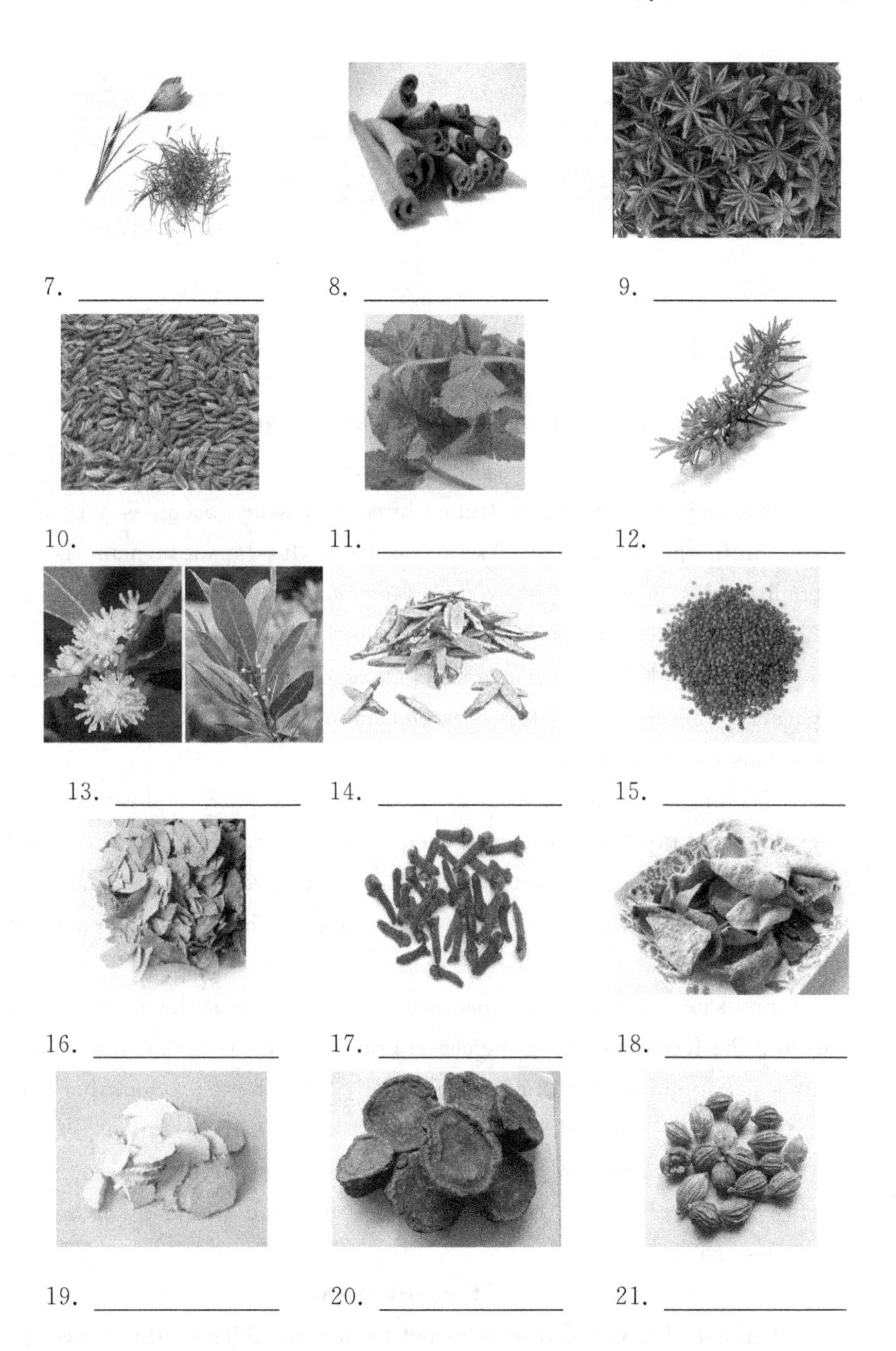
7. ______
8. ______
9. ______
10. ______
11. ______
12. ______
13. ______
14. ______
15. ______
16. ______
17. ______
18. ______
19. ______
20. ______
21. ______

22. ______________ 23. ______________ 24. ______________

Activity 2

Task Read the following passages and answer the questions.

Soy Sauce

A condiment originally from China, soy sauce occupies a preeminent position in the cuisines of Asian countries. Its Japanese name is shoyu. Traditionally, soy sauce, shoyu and tamari refer to the liquid formed during the manufacture of miso.

Traditional Chinese soy sauce is made using whole soybeans and ground wheat. It can be more or less dark depending on its age and whether caramel or molasses has been added.

Tamari is made exclusively using soybeans or soybean meal (the residue from pressing the beans when oil is extracted); therefore, it contains no cereal grain. It sometimes contains additives such as monosodium glutamate and caramel. Tamari is dark and has a thicker consistency. Shoyu is lighter in color than Chinese soy sauce and slightly sweet.

Soy sauce (Chinese or Japanese) contains some of the alcohol produced during the fermentation of the cereal grains, whereas tamari has none. The soy sauce found in supermarkets is usually a synthetic product that is a pale imitation of the original.

Question 1: Which part of the world is soy sauce often used?

Question 2: What is the difference between Chinese soy sauce and Japanese shoyu?

Culinary Herbs

Culinary herbs are distinguished from vegetables in that, like spices, they are used in small amounts and provide flavor rather than substance to food. Many culinary herbs are perennials such as thyme or lavender, while others are biennials such as parsley or annuals like basil. Some perennial herbs are shrubs (such as rosemary, Rosmarinus officinalis), or trees (such as bay laurel, Laurus nobilis)—this contrasts with botanical herbs, which by definition cannot be woody plants. Some plants are used as both herbs and spices, such as dill weed and dill seed or coriander leaves and seeds. Also, there are some herbs such as those in the mint family that are used for both culinary and medicinal purposes.

Question 1: How many kinds of herbs are presented in the article?

Question 2: What kinds of herbs can be used as both herbs and spices?

Activity 3

Task 1 Translate the following sentences into English.

1. 鱼露是沿海地区的传统发酵调味品。

____

2. 然后你把佐料放进面条里,再加点盐。

____

3. 日本芥末通常配生鱼片。

____

4. 这肉里应该用些盐和芥末调味。

____

5. 在法国,芥菜籽被浸透,然后再磨成糨糊状。

____

6. 酸辣酱可以混合在任何一种印度餐中,形成一种不同的口味。

____

7. 我想在上面放一些火腿、香肠、蘑菇、洋葱、橄榄和菠萝。

____

8. 盐是一种常用的食物防腐剂。

____

9. 肉豆蔻常用作食物中的香料。

____

10. 这个汤的特殊香味是因为藏红花粉。

____

Task 2 Translate the following passage into English.

对于中国人来说,肉豆蔻作为香料的功能更为普通人所熟知。在西方,欧洲人有许多使用肉豆蔻的方法。英国人把肉豆蔻加在米糕、蛋挞和牛乳中;法国人常在糕饼、肉馅饼和香肠中放入肉豆蔻调味;意大利人,则特别钟爱加入肉豆蔻酱汁的小牛排;荷兰人做的炖菜料理,更少不了肉豆蔻。

Unit 4 Meat

Learning goals

To know essential information of various kinds of meat and the usage of them in cooking.

Vocabulary

flank [flæŋk] n. 侧面
brisket [ˈbrɪskɪt] n. 胸部
shank [ʃæŋk] n. 胫
rump roast 烤蹄髈,烤大腿部分
bacon [ˈbeɪkən] n. 培根
ham [hæm] n. 火腿
sausage [ˈsɒsɪdʒ] n. 香肠
goose foie gras 鹅肝酱
turkey [tɜːki] n. 火鸡
hen [hen] n. 母鸡
capon [ˈkeˌpɒn, -pən] n. 阉(公)鸡
pigeon [ˈpɪdʒən] n. 鸽子
lamb [læm] n. 羔羊
veal [viːl] n. 小牛肉

venison [ˈvenɪsən, -zən] *n.* 鹿肉
ground meat 肉馅
blood sausage 血肠
skirt steak 肋排
eye round roast 烤眼肉
top round roast 烤里仔盖
T-bone steak 丁骨牛排
boneless top loin steak 腰脊牛排
tenderloin roast 烤菲力
porterhouse steak 上等腰肉牛排
back ribs 肋骨小排
top sirloin steak 沙朗牛排,菲力牛排
chuck eye roast 前肩肉眼/牛排
blade roast 煎颈片肉
smoked ham 熏火腿
pancetta [pænˈsetə] *n.* 未经熏制的咸肉
prosciutto [prəʊˈʃuːtəʊ] *n.* 意大利熏火腿
Frankfurt sausage 法兰克福香肠
Genoa salami 热那亚式萨拉米香肠(无烟熏制的猪肉香肠)
kielbasa [kɪlˈbɑːsə] *n.* 波兰熏肠
chorizo [tʃəˈriːzəu] *n.* 西班牙加调料的口利左香肠
mortadella [ˌmɔːtəˈdelə] *n.* 熏香肠
bratwurst [ˈbrætwɜːst] *n.*(供煎食的)德国式小香肠
Chinese dried sausage 中式香肠
German salami 德国香肠
Toulouse sausage 图卢兹香肠(肥猪肉生香肠)
pepperoni [ˌpepəˈrəuni] *n.* 意大利辣香肠

Dialogue 1

A: What's this?

B: It's a colander.

A: How can we use it?

B：It can be used as a strainer for draining vegetables and fruits.

A：Can it be used for washing as well?

B：Definitely.

A：What shall we do then?

B：We defrost the fish with the microwave first.

A：I see. What shall we use，a pan or a wok to fry fish?

B：Pans are not as handy as woks for Chinese cooking. Could you do me a favor to make sorbet with the blender?

A：My pleasure.

Dialogue 2

(*J=Jack*，*A=Andy.*)

J：Good morning，Andy.

A：Good morning，Jack.

J：You look great!

A：Thank you.

J：What a nice day today! Let's start work.

A：Wait a minute，Jack. Let me cover my cut with bandage.

J：OK，hurry up!

Activity 1

Task　Try to write the names in English below the pictures.

1. ________　2. ________　3. ________

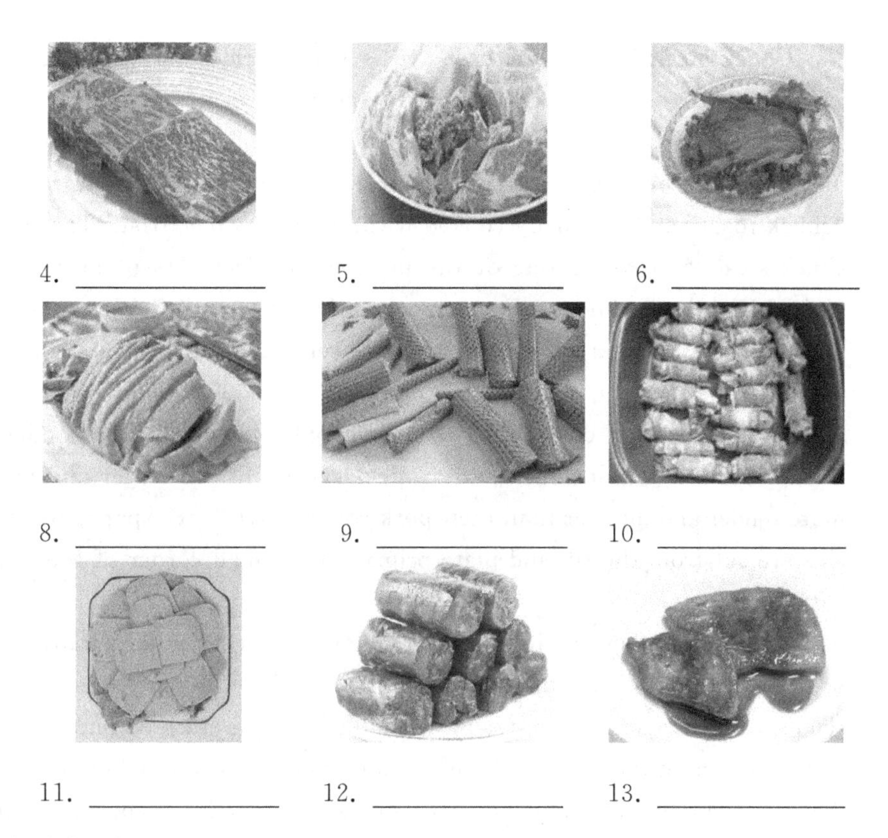

4. ________ 5. ________ 6. ________

8. ________ 9. ________ 10. ________

11. ________ 12. ________ 13. ________

Activity 2

Task Read the following passage based on the picture; discuss with your partner about the usage of every part of beef in cooking.

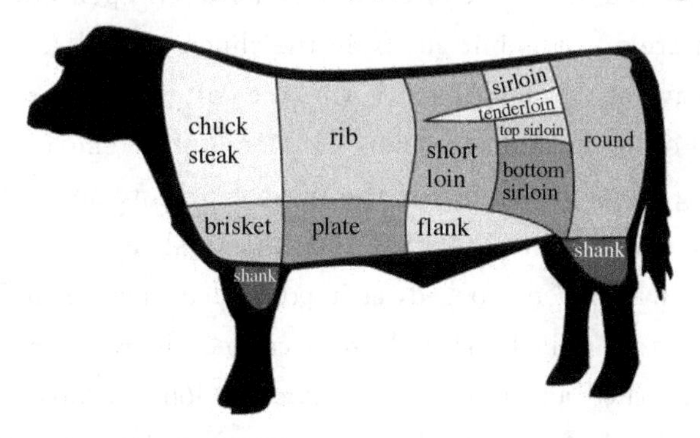

Chuck steak is a cut of beef and is part of the sub primal cut. The typical chuck steak is a rectangular cut, about 1 "thick and containing parts of the shoulder bones", and is often known as a "7-bone steak". This cut is usually grilled or broiled; a thicker version is sold as a "7-bone roast" or "chuck roast" and is usually cooked with liquid as a pot roast. The bone-in chuck steak or roast is one of the more economical cuts of beef. In the United Kingdom, this part is commonly referred to as "braising steak". It is particularly popular for use as ground beef, due to its richness of flavor and balance of meat and fat.

Short ribs (UK cut: thin rib) (commonly known in UK as "Jacob's Ladder") are a popular cut of beef. Beef short ribs are larger and usually more tender and meatier than their pork counterpart, pork spare ribs. Short ribs are cut from the rib and plate primal and a small corner of the square-cut chuck.

A full slab of short ribs is typically about 10 inches square, ranges from 3—5 inches thick, and contains three or four ribs, intercostal muscles and tendon, and a layer of boneless meat and fat which is thick on one end of the slab and thin on the other. There are numerous ways to butcher short ribs. The ribs can be separated and cut into short lengths (typically about 2 inches long), called an "English cut"; "flanken cut" across the bones (typically about 1/2 inch thick); or cut into boneless steaks (however, these are not to be confused with "boneless country-style short ribs", a cut recently introduced in the United States as a cheaper alternative to rib steak, which are not ribs but cut from the chuck eye roll).

In Korea, short lengths of rib are often further butchered by butterflying (or using an accordion cut) to unfurl the meat into a long ribbon trailing from the bone, or the meat can be removed from the bone entirely and cut into thin (1/4—1/8 inch thick) slices.

Short ribs may be long-cooked, as in pot-au-feu, a classic of French cuisine, or rapidly seared or grilled, as in Korean cuisine, in which short ribs (called galbi), are marinated and grilled over charcoal (long-cooking thinner, shorter cuts are also a Korean favorite). A specific type from Hawaii is known as Mauistyle ribs. Flanken is a traditional Eastern European Jewish dish, and the origin of that name for short ribs cut across the bone. Flanken are boiled in broth with onions and other seasonings, until the meat is very tender and the broth is rich, and served with grated horseradish. Other popular preparations are barbecue and braising. A highly regarded recent method is to cook them sous-vide for up to 72 hours.

Brisket is a cut of meat from the breast or lower chest of beef or veal. The beef brisket is one of the nine beef prime cuts. The brisket muscles include the superficial and deep pectorals. As cattle do not have collar bones, these muscles support about 60% of the body weight of standing/moving cattle. This requires a significant amount of connective tissue, so the resulting meat must be cooked correctly to tenderize the connective tissue.

According to the *Random House Dictionary of the English Language, Second Edition*, the term derives from the Middle English brisket which comes from the earlier Old Norse brjósk, meaning cartilage. The cut overlies the sternum, ribs and connecting costal cartilages.

Short loin is a cut of beef that comes from the back of the steer or heifer. It contains part of the spine and includes the top loin and the tenderloin. This cut yields types of steak including porterhouse, strip steak (Kansas City Strip, New York Strip), and T-bone (a cut also containing partial meat from the tenderloin). The T-bone is a cut that contains less of the tenderloin than does the porterhouse. *Webster's Dictionary* defines it as "a portion of the hindquarter of beef immediately behind the ribs that is usually cut into steaks."

The sirloin steak is a steak cut from the rear back portion of the animal, continuing off the short loin from which T-bone, porterhouse, and club steaks are cut.

The sirloin is actually divided into several types of steak. The top sirloin is the most prized of these and is specifically marked for sale under that name. The bottom sirloin, which is less tender and much larger, is typically marked for sale simply as "sirloin steak." The bottom sirloin in turn connects to the sirloin tip roast.

In British and Australian butchery, the word sirloin refers to cuts of meat from the upper middle of the animal, similar to the American short loin.

A round steak is a steak from the round primal cut of beef. Specifically, a round steak is the eye (of) round, bottom round, and top round still connected, with or without the "round" bone (femur), and may include the knuckle (sirloin tip), depending on how the round is separated from the loin. This is a lean cut and it is moderately tough. Lack of fat and marbling makes round dry out when cooked with dry-heat cooking methods like roasting or grilling. Round steak is commonly prepared with slow moist-heat methods including braising, to tenderize the meat and maintain moisture. The cut is often sliced thin, then dried or smoked at low temperature to make jerky.

Rump cover, with its thick layer of accompanying fat, is considered one of the best (and most flavorful) beef cuts in many South American countries, particularly Brazil and Argentina. This specific cut does not tend to be found elsewhere, however.

The flank steak is a beef steak cut from the abdominal muscles of the cow. A relatively long and flat cut, flank steak is used in a variety of dishes including London broil and as an alternative to the traditional skirt steak in fajitas. It can be grilled, pan-fried, broiled, or braised for increased tenderness.

The cut is common in Colombia, where it is known as sobrebarriga, literally meaning "over the belly".

Flank steak is best when it has a bright red color. Because it comes from a strong, well-exercised part of the cow, it is best sliced across the grain before serving.

Flank steak is frequently used in Asian cuisine, often sold in Chinese markets as "stir-fry beef". In the United Kingdom McDonald's uses "100% forequarter and flank" in its burger patties.

The beef shank is the shank (or leg) portion of a steer or heifer. In Britain the corresponding cuts of beef are the shin (the foreshank), and the leg (the hindshank).

Due to the constant use of this muscle by the animal it tends to be tough, dry, and sinewy, so is best when cooked for a long time in moist heat. It is an ideal cut to use for beef bourguignon. As it is very lean, it is widely used to prepare very low-fat ground beef. Due to its lack of sales, it is not often seen in shops. Although, if found in retail, it is very cheap and a low-cost ingredient for beef stock. Beef shank is a common ingredient in soups.

In Australia, it is commonly sold from butchers as gravy beef.

Activity 3

Task 1 Translate the following sentences into English.

1. 我要吃水芹菜汤和牛排。

____

2. 打三个蛋，再加盐。

____

3. 他把小牛肉切成肉片。

____

4. 他把一片无骨牛排放进小圆面包里。

____

5. 切一片牛的下部胸肉。

____

6. 将煎好的牛胸肉移放到一个大烤盘里，较肥面朝上。

____

7. 这是从牛的腰部嫩肉切下来的无骨肉排。

____

Task 2　Translate the following article into English.

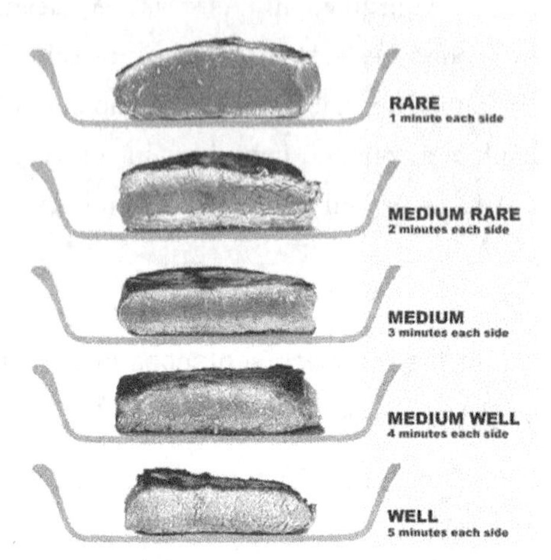

Rare, very rare:极生,煎的(grill)时间不超过 3 分钟。外表有烧烤过的痕迹,但是里面还是冷得几乎没有受到热度。切开时还有血水渗出,但是肉质极嫩,口感多汁。

Rare:生,煎的时间不超过 4 分钟。外表有烤焦痕迹,里面肉质呈现原来红色,但入口有热度。切开时还有血水渗出,但是肉质极嫩,口感多汁。

Medium rare:中生,煎的时间 6—8 分钟。外表有烧烤过的痕迹,里面已经全面加热,可以感受到相当热度,但是肉质还是呈现红色。切开时还有少许血水渗出,但是肉质嫩,口感多汁。

Medium:稍熟,通常说的 5－6 分熟。煎的时间 8—10 分钟。外表烧烤呈深褐色,但是里面除了中间部分呈现粉红色外,外围部分呈现烧烤过的浅褐色。切开时流出褐色肉汁,需要咬上数口才能咽下。

Medium well:中熟,7 分熟。煎的时间 10—12 分钟。外表烧烤呈深褐色,但是除里面核心部分呈现少许红色外,外围部分呈现烧烤过的褐色。切开时流出褐色肉汁,需要咬上数口才能咽下。

Well done:全熟,煎的时间 12—15 分钟。外表已有明显烤焦痕迹,热度已经渗入整片肉,里面肉色因为高度加热呈现深褐色。咬劲很够才能下咽。

Activity 4

Task 1 Topic discussion.

What do you think about fast food, such as high-calorie and time-saving?

In what case will you choose fast food? Any problems caused by fast food?

Task 2 Further reading.

Types of Dairy Products

- Crème fraîche, slightly fermented cream
- Clotted cream, thick, spoonable cream made by heating
- Smetana, Central and Eastern European variety of sour cream
- Cultured milk resembling buttermilk, but uses different yeast and bacterial cultures
- Kefir, fermented milk drink from the Northern Caucasus
- Kumis/airag, slightly fermented mares' milk popular in Central Asia
- Powdered milk (or milk powder), produced by removing the water from (usually skim) milk
- Condensed milk, milk which has been concentrated by evaporation, with sugar added for reduced process time and longer life in an opened can
- Evaporated milk, (less concentrated than condensed) milk without added sugar
- Ricotta, acidified whey, reduced in volume
- Khoa, dairy product used in Indian cuisine
- Baked milk, a variety of boiled milk that has been particularly popular in Russia
- Butter, mostly milk fat, produced by churning cream
- Buttermilk, the liquid left over after producing butter from cream, often dried as livestock feed
- Ghee, clarified butter, by gentle heating of butter and removal of

the solid matter

- Smen, a fermented, clarified butter used in Moroccan cooking
- Cheese, produced by coagulating milk, separating from whey and letting it ripen, generally with bacteria and sometimes also with certain molds
- Curds, the soft, curdled part of milk (or skim milk) used to make cheese
- Whey, the liquid drained from curds and used for further processing or as a livestock feed
- Cream cheese, produced by the addition of cream to milk and then curdled to form a rich curd or cheese
- Yogurt, milk fermented by Streptococcus salivarius ssp.
- Clabber, milk naturally fermented to a yogurt-like state
- Gelato, slowly frozen milk and water, lesser fat than ice cream
- Ice cream, slowly frozen cream, milk, flavors and emulsifying additives
- Ice milk, low-fat version of ice cream
- Frozen custard
- Frozen yogurt, yogurt with emulsifier

Century Egg

Century egg or pidan (皮蛋), also known as preserved egg, hundred-year egg, thousand-year egg, thousand-year-old egg, and millennium egg, is a Chinese cuisine ingredient made by preserving duck, chicken or quail eggs in a mixture of clay, ash, salt, quicklime, and rice hulls for several weeks to several months, depending on the method of processing.

Through the process, the yolk becomes a dark green to grey colour, with a creamy consistency and an odor of sulphur and ammonia, while the white becomes a dark brown, translucent jelly with little flavor. The transforming agent in the century egg is its alkaline material, which gradually raises the pH of the egg to around 9, 12, or more during the curing process. This chemical process breaks down some of the complex, flavorless proteins and fats, which produces a variety of smaller flavorful compounds.

Some eggs have patterns near the surface of the egg white that are likened to pine branches, and that gives rise to one of its Chinese names, the pine-patterned egg.

Bacon

Bacon is a cured meat prepared from a pig. It is first cured using large quantities of salt, either in a brine or in a dry packing; the result is fresh bacon (also known as green bacon). Fresh bacon may then be further dried for weeks or months in cold air, or it may be boiled or smoked. Fresh and dried bacon is typically cooked before eating. Boiled bacon is ready to eat, as is some smoked bacon, but may be cooked further before eating.

Bacon is prepared from several different cuts of meat. It is usually made from side and back cuts of pork, except in the United States, where it is almost always prepared from pork belly (typically referred to as "streaky" "fatty" or "American style" outside of the US and Canada). The side cut has more meat and less fat than the belly. Bacon may be prepared from either of the two distinct back cuts: fatback, which is almost pure fat, and pork loin, which is very lean. Bacon-cured pork loin is known as back bacon.

Uncooked pork belly bacon strips. Bacon may be eaten smoked, boiled, fried, baked, or grilled, or used as a minor ingredient to flavor dishes. Bacon is also used for barding and larding roasts, especially game, e. g. venison or pheasant. The word is derived from the Old High German bacho, meaning "buttock" "ham" or "side of bacon", and cognate with the Old French bacon.

In continental Europe, this part of the pig is usually not smoked like bacon in the United States; it is used primarily in cubes (lardons) as a cooking ingredient, valued both as a source of fat and for its flavor. In Italy, this is called pancetta and is usually cooked in small cubes or served uncooked and thinly sliced as part of an antipasto.

Meat from other animals, such as beef, lamb, chicken, goat, or turkey, may also be cut, cured, or otherwise prepared to resemble bacon, and may even be referred to as "bacon". Such use is common in areas with significant Jewish and Muslim populations. The USDA defines bacon as "the cured belly of a swine carcass"; other cuts and characteristics must be separately qualified (e. g. "smoked pork loin bacon"). For safety, bacon must be treated to prevent trichinosis, caused by Trichinella, a parasitic roundworm which can be destroyed by heating, freezing, drying, or smoking.

Bacon is distinguished from salt pork and ham by differences in the brine (or dry packing). Bacon brine has added curing ingredients, most notably sodium nitrite, and occasionally sodium nitrate or potassium nitrate (saltpeter); sodium ascorbate or erythorbate are added to accelerate curing and stabilize color. Flavorings such as brown sugar or maple are used for some products. If used, sodium polyphosphates are added to improve sliceability and reduce spattering when the bacon is pan fried. Today, a brine for ham, but not bacon, includes a large amount of sugar. Historically, "ham" and "bacon" referred to different cuts of meat that were brined or packed identically, often together in the same barrel.

Unit 5 Seafood

Learning goals

To know essential information of various kinds of seafood and the usage of them in cooking.

Vocabulary

eel [il,iːl] *n.* 鳗鱼
bass [beɪs] *n.* 鲈鱼
pike [paɪk] *n.* 长矛;梭鱼
carp [kɑːp] *n.* 鲤鱼
pike-perch *n.* 狮鲈;梭鲈
trout [traʊt] *n.* 鲑鳟鱼

mullet [ˈmʌlɪt] *n.* 胭脂鱼；鲻鱼
bluefish [ˈbluːˌfɪʃ] *n.* 竹荚鱼类
shad [ʃæd] *n.* 西鲱，美洲西鲱
monkfish [ˈmʌŋkfɪʃ] *n.* 安康鱼
sea bass 〈美〉黑鲈，鲈科鱼的总称
sturgeon [ˈstɜːdʒən] *n.* 鲟
caviar [ˈkæviˌɑː(r)] *n.* 鱼子酱
sardine [ˌsaːˈdiːn] *n.* 沙丁鱼
anchovy [æntʃəvi] *n.* 凤尾鱼；鳀鱼
cod [kɒd] *n.* 鳕鱼
salmon [ˈsæmən] *n.* 鲑鱼，大马哈鱼；鲑鱼肉
tuna [ˈtjuːnə] *n.* 〈鱼〉金枪鱼(科)，鲔鱼；金枪鱼罐头
shark [ʃɑːk] *n.* 鲨鱼
sole [sol] *n.* 鳎(可食用比目鱼)
scallop [ˈskɒləp] *n.* 扇贝；扇贝壳；扇(贝)形
mussel [ˈmʌsəl] *n.* 贻贝，蚌类；淡菜
oyster [ˈɒɪstə(r)] *n.* 牡蛎
octopus [ˈɒktəpəs] *n.* 章鱼
cuttlefish [ˈkʌtlfɪʃ] *n.* 乌贼，墨鱼
abalone [ˌæbəˈləuni] *n.* 〈美〉鲍鱼
shrimp [ʃrɪmp] *n.* 虾，小虾
crab [kræb] *n.* 蟹，蟹肉
lobster [ˈlɒbstə] *n.* 龙虾；龙虾肉
swordfish [ˈsɔːdfɪʃ] *n.* 旗鱼
haddock [ˈhædək] *n.* 小口鳕，黑线鳕(产于北大西洋的食用鱼)
black pollock 黑鳕鱼
whiting [waɪtɪŋ] *n.* 牙鳕；石首鱼
turbot [ˈtɜːbət] *n.* 大菱鲆
pink shrimp 桃红对虾
giant tiger shrimp 巨虎虾
scampi [ˈskæmpi] *n.* 挪威海螯虾
crayfish [ˈkreɪˌfɪʃ] *n.* 淡水螯虾(肉)；龙虾

Dialogue 1

Directions: Practice the following dialogue in pairs.

(*W*=*Waiter*, *M1*=*Mark*, *M2*=*Marian*.)

W: Would you like to order now, sir?

M1: Yes, I think so. Marian?

M2: Yes, I'll have the salmon teriyaki, please.

W: And what kind of potatoes would you like to go with that?

M2: Baked, please. For the vegetable, I'd like broccoli.

W: And would you care for soup or salad to start with?

M2: I think I'll have a salad, please.

W: All right. With what kind of dressing?

M2: I'd like blue cheese.

W: Yes. And you, sir? What will you have?

M1: Those lobster tails on this menu sound pretty good.

W: Oh, I'm very sorry, sir. We don't have any lobster now.

M: No lobster? OK…I guess I'll take the steak then. Rare.

W: Yes. What about potatoes? Mashed, boiled or baked?

M1: Mashed potatoes. For vegetable, I'd like asparagus.

W: And, soup or salad?

M1: Oh, I'll try the cream of cauliflower.

W: Good. Anything to drink while you wait?

M2: Iced water, please.

M: Make that two.

Dialogue 2

Directions: Practice the following dialogue in pairs.

(*L*=*Lisa*, *S*=*Sherry*.)

L: What kind of pasta do you want?

S: We can try the spinach pasta. Maybe it's healthier.

L: I hope Nigel doesn't complain.

S: Don't worry. He's not finicky. We still need to get tomato sauce.

L: Right. Does he like spaghetti or…?

S: He likes spaghetti, He likes Chinese noodles, too.

L: I know a good brand, Paul Newman's. I hope they still sell it.

S: So where is the tomato sauce?

L: It's over there.

S: Let's see. Beef, onions, garlic, pasta, tomato sauce. What else do we need?

L: Lettuce for salad.

S: What kind of salad will you make?

L: Caesar salad.

Dialogue 3

Directions: Practice the following dialogue in pairs.

(*S=Steven*, *G=George*.)

S: George, do you know how to scale fish?

G: Tell me, Steven.

S: OK, most fish have scales that must be removed before cooking.

G: Yes, I know that. But how?

S: The best way to remove scales is with a fish scaler.

G: Are there any other tools that can be used for scaling fish?

S: Yes, the dull side of a knife, or a spoon handle also can be used if a scaler is not available.

G: To scale a fish, should I start from the tail or from the head?

S: You should work from the tail toward the head.

G: It is hard to grip(紧抓) it.

S: Yes. Grip the fish by the tail, and allow water to flow over the fish to help keep the scales from flying around (飞来飞去).

G: Like this?

S: Very good, George! Do not pinch the fish too tightly as this could bruise the flesh.

G: Got it, thank you.

S: Enjoy your work!

Activity 1

Task Try to write the names in English below the pictures.

13. ______________ 14. ______________

Task 2 Translate the following sentences into English.

1. 鳕鱼已用盐腌起留着日后吃。

____

2. 黑线鳕通常烘焙,但有时涂大量黄油后烧烤。

____

3. 人们常常把大西洋鳕鱼或黑线鳕去骨切片后烹调。

____

4. 日本为鲔鱼主要消费国之一。

____

5. 先生,这就是你点的清煎鲤鱼和烤牛肉。

____

6. 安康鱼配上橙子感觉非常精巧美妙。

____

7. 日本人喜欢吃生的鲑鱼肉。

____

8. 可食用的金枪鱼肉通常是装在罐头中或加工过的。

____

9. 扇贝洗净沥干水,每只开半,然后再切成小片。

____

Activity 2

Task Reading.

Seafood

Seafood is any form of sea life regarded as food by humans. Seafood prominently includes fish and shellfish. Shellfish include various species of molluscs, crustaceans, and echinoderms. Historically, sea mammals such as whales and dolphins have been consumed as food, though that happens to a lesser extent these days. Edible sea plants, such as some seaweeds and microalgae, are widely eaten as seafood around the world, especially in Asia (see the category of sea vegetables). In North America, although not generally in the United Kingdom, the term "seafood" is extended to fresh water organisms eaten by humans, so all edible aquatic life may be referred to as seafood. For the sake of completeness, this article includes all edible aquatic life.

The harvesting of wild seafood is known as fishing, and the cultivation and farming of seafood is known as aquaculture, mariculture, or in the case of fish, fish farming. Seafood is often distinguished from meat, although it is still animal and is excluded in a strict vegetarian diet. Seafood is an important source of protein in many diets around the world, especially in coastal areas.

Most of the seafood harvest is consumed by humans, but a significant proportion is used as fish food to farm other fish or rear farm animals. Some seafoods (kelp) are used as food for other plants (fertilizer). In these ways, seafoods are indirectly used to produce further food for human consumption. Products, such as fish oil and spirulina tablets are also extracted from seafoods. Some seafood is feed to aquarium fish, or used to feed domestic pets, such as cats, and a small proportion is used in medicine, or is used industrially for non-food purposes (leather).

Chapter 5

Recipes

Unit 1 Hors d'oeuvres

Learning goals

To know essential information of various kinds of hors d'oeuvres, including canapés, caviar and cold cuts.

To get to know the making process of some classic hors d'oeuvres.

Vocabulary

hors d'oeuvre 餐前点心;拼盘
cocktail [ˈkɒkteɪl] *n.* 鸡尾酒;餐前开胃菜
appetizer [ˈæpɪtaɪzə(r)] *n.* 开胃品
relish [ˈrelɪʃ] *n.* 味道,滋味;调味品
skewer [skjuːə(r)] *n.* 串肉杆;烤肉叉子
caviar [ˈkæviɑː(r)] *n.* 鱼子酱
dip [dɪp] *n.* 蘸
cold cuts 冷盘
deviled eggs 煮鸡蛋
dumpling [ˈdʌmplɪŋ] *n.* 饺子
bruschetta [brʊˈsketə] *n.* 意大利烤面包片
tongue toast 牛舌多士

California fusion peach salsa 加利福尼亚桃酱
crab quesadillas with mango salsa 蟹炸玉米饼配芒果沙拉
cream cheese wontons 奶油芝士馄饨
deep fried pretzels 油炸椒盐脆饼
fresh mozzarella bruschetta 新鲜的意大利干酪面包
guacamole [ˌgwækəˈməuli] n. 鳄梨酱；鳄梨色拉
hornickels' glazed onion canapes 琉璃洋葱开胃菜
hummus with sprouted chickpea 发芽鹰嘴豆泥
quesadilla [ˌkeɪsəˈdiːjə] n. 油炸玉米粉饼(一种墨西哥食品,以干酪等为馅)
easy nachos 玉米片
red and white onion tapenade 红色和白色的洋葱酱
roasted chicken nachos with green chili-cheese sauce 烤玉米片和绿辣椒奶酪酱
San Francisco style scallop ceviche 三藩风格扇贝沙拉
spicy chickpeas 香辣鹰嘴豆
spinach artichoke dip 菊芋浸菠菜
steamed artichoke 蒸菊芋
toasted baguette slices with pecan, butter and brie 核桃、黄油、奶酪烤面包片
ham and cheese crepes 火腿和奶酪薄饼

Dialogue

(*C1*=*Commis*, *C2*=*Chef*.)
C1: What should I do?
C2: Make some eggplant dip.
C1: What should I do first?
C2: Get 15 eggplants.
C1: What is next?
C2: Prick the eggplants with a fork.
C1: I did that.
C2: Put the eggplants in the oven, please.
C1: At what temperature?

C2：At 400℃.

C1：For how long?

C2：40 minutes.

C1：And in the meantime?

C2：Cut the onions into fourths.

C1：OK，I am finished.

C2：Good. Next，peel ten cloves of garlic.

C1：I've already peeled the garlic.

C2：We also need lemon juice. Please squeeze some lemons until you have two and a half cups. Use a measuring cup.

C1：OK.

C2：Put the eggplants，onions，garlic，lemon juice，and 10 tablespoons of olive oil in the blender.

C1：Should I add some salt?

C2：Yes，1 tablespoon of salt and blend at high speed.

Activity 1

Task 1 Canapé Smoked Salmon and Passion Fruit Roulade (肉卷).

Basic recipe：10 servings.

Ingredients：

600 g smoked salmon

4 ea sheet gelatin

250 ml fresh cream

100 g sugar

150 ml passion fruit puree (百香果酱)

Caviarchives

Directions：

Bring the passion fruit and sugar to boil and reduce by half; slice the smoked salmon into long thin slices.

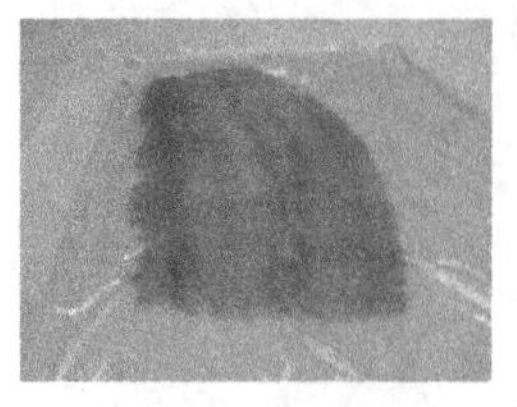

Lay out the sliced salmon on a plastic film, 3 slices side by side. Whip the cream and add the reduced cold passion fruit coulis.

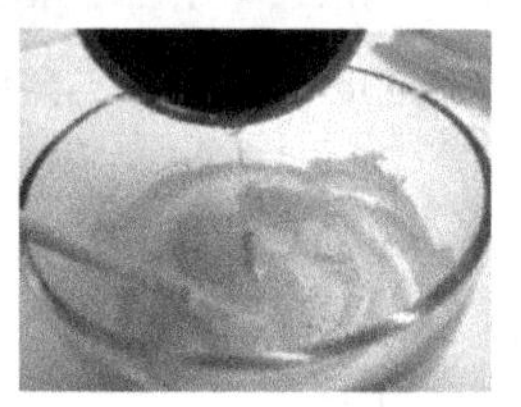
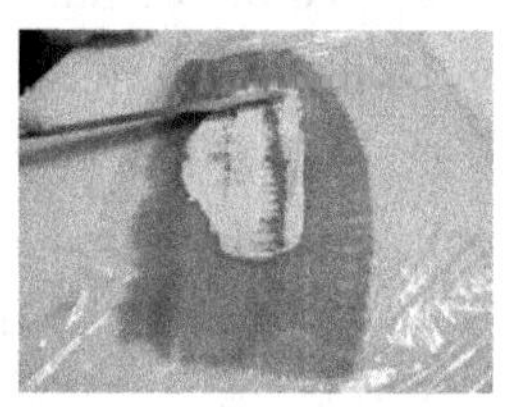

Mix gently and add the dissolved gelatin. Spread the mixture evenly onto the salmon.

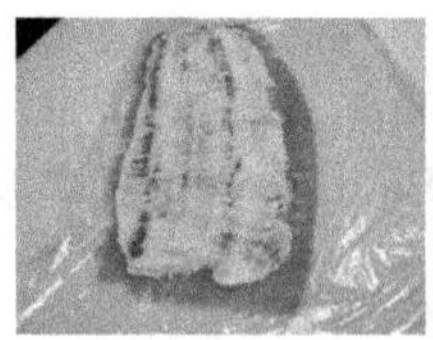
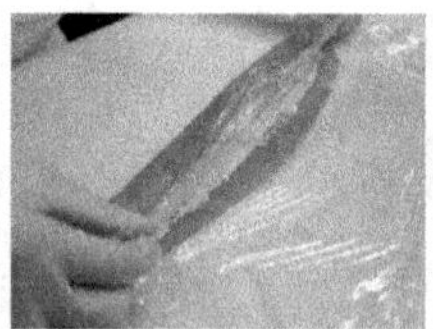
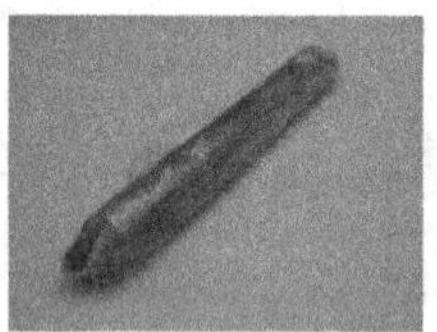
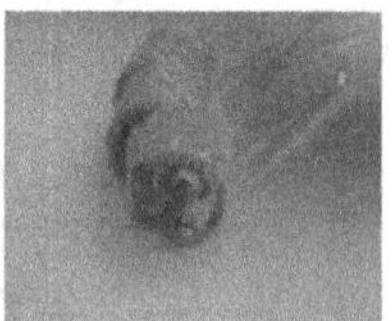

Cut the roulade with the foil into 4 cm pieces and place them on a small plate or platter, garnish with sour crème, caviar and chives.

Task 2 Smoked Salmon with Dill Sour Cream and Asparagus.

Ingredients:

Smoked salmon

French baguette

Lettuce leaves

Thai asparagus

Butter fresh

Dill

Sour cream

Salt and pepper

Directions:

Cut the baguette in a slight angle 1.5 cm thick, butter the slices evenly, and put on some lettuce.

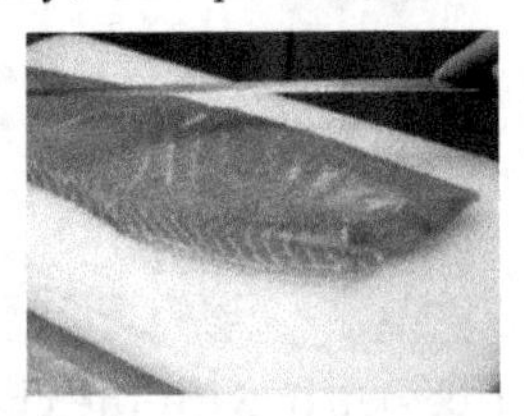

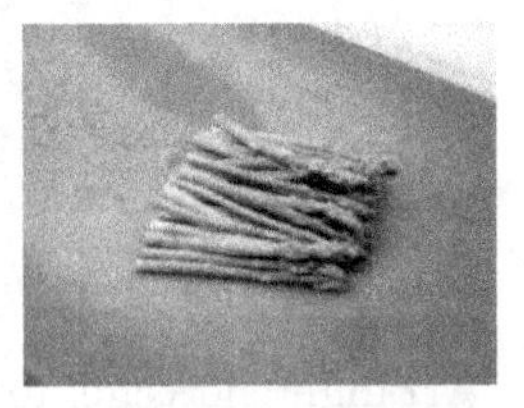

Clean off the dry skin of the salmon and slice salmon (if no pre-sliced salmon available), put salmon slices on top of the lettuce, trim the asparagus and blanch quickly, refresh in ice water to maintain green color, trim to required length and put 3 pieces on top of the salmon.

Season with pepper mill, make a sauce out of sour cream and chopped dill, season to taste and spoon on top of the asparagus, before serving.

Task 3 Canapé Vietnamese Vegetarian Spring Roll.

Ingredients:

3 kinds of capsicum

Coriander

Chili

Rice paper (糯米纸)

Sweet chili sauce

Directions:

Cut capsicums and chili in julienne, pick the coriander, soak the rice paper in cold water, when gently put on clean towel.

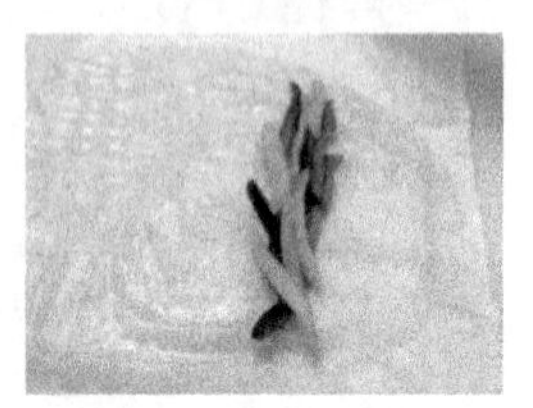 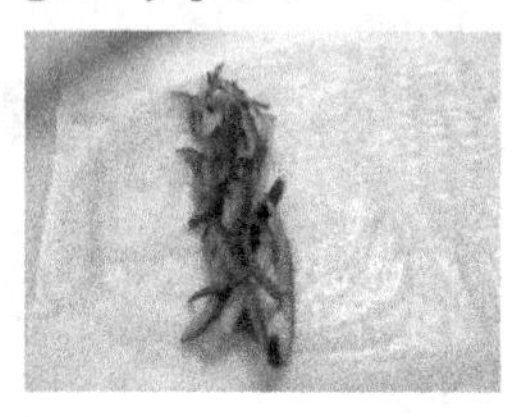 

Arrange the vegetables lengthwise onto the rice paper with the chili and coriander, season.

 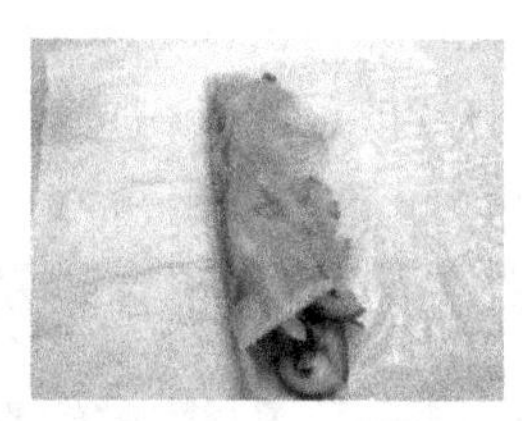 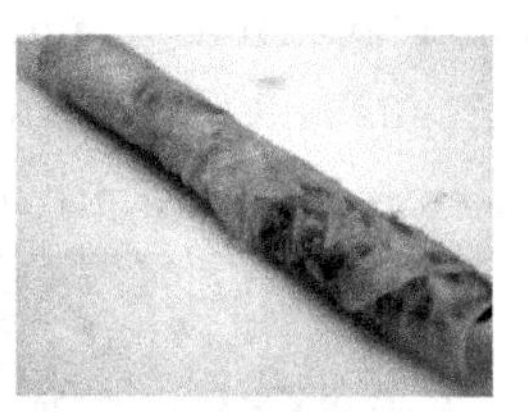

Roll tight to a roulade, and ensure not rolling the roulade to thick, so it will fit in the glass for serving.

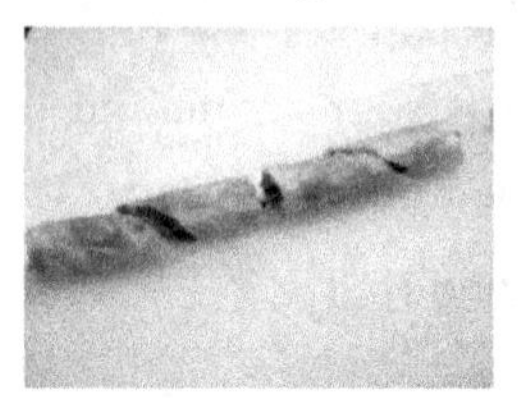 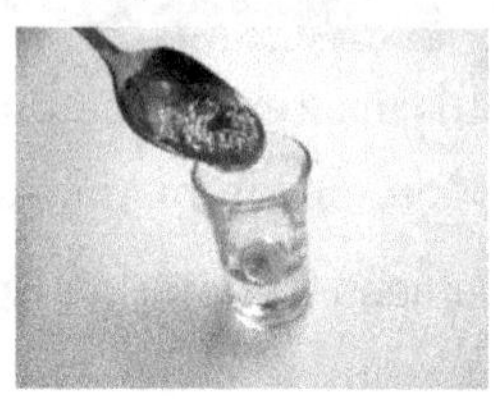 

Cut into 6 cm long pieces, fill some sweet chili sauce into a shooter glass and put on piece of the roll in the glass garnish with frisee lettuce and chives.

Task 4 Prawn and Mango Skewer with Gazpacho Sauce.

Ingredients:

Gazpacho sauce (西班牙凉菜酱)

Prawns small to medium size

Mangoes

Wooden skewers

Directions:

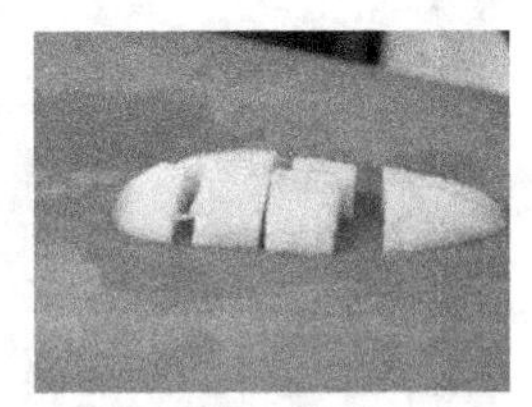

Peel mango and cut 1 to 1.5 cm thick cubes.

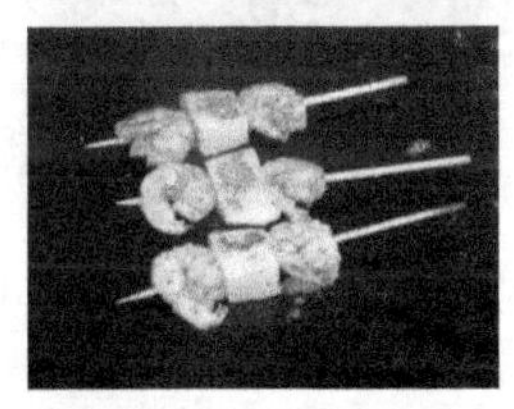

Skewer shrimps and mango on a wooden skewer or lemon grass and cook quickly on a flat top grill.

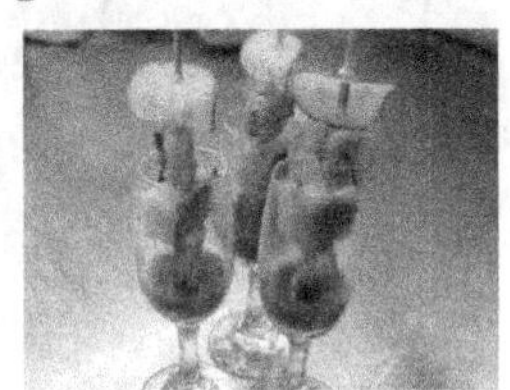

Fill shooter glasses or sherry glasses to a quarter with gazpacho sauce, garnish the skewer with a cucumber slice and put into the glass.

Activity 2

Write down the cooking process of the following dishes according to ingredients given below and number the pictures.

Task 1　Canapé Melon with Parma Ham.

Ingredients:

Sweet melon

Parma ham (帕尔玛火腿)

Sun dried tomato (风干番茄)

Chervil (细叶芹)

Task 2　Canapé Caviar.

Ingredients:

Eggs boiled

Chives chopped

Caviar (your choice)

Sour crème

Onions, red, finely diced

Plastic ring (塑料盛器)

Activity 3

Task 1 Reading.

Hors d'oeuvre

Hors d'oeuvres, literally apart from the first course, are food items served before the main courses of a meal.

If there is an extended period between when guests arrive and when the meal is eaten (for example, during a cocktail hour), these might also serve the purpose of sustaining guests during the wait, in the same way that apéritifs are served before meals. Hors d'oeuvres are sometimes served with no meal afterward. This is the case with many reception and cocktail party events.

Hors d'oeuvre may be served at the dinner table as a part of the meal, or they may be served before seating. Hors d'oeuvres prior to a meal are either rotated by waiters or passed. Stationary hors d'oeuvres served at the table may be referred to as "table hors d'oeuvre". Passed hors d'oeuvres may be referred to as "butler-style" or "butlered" hors d'oeuvre.

Though any food served prior to the main course is technically an hors d'oeuvre, the phrase is generally limited to individual items, not crudités, cheese or fruit. For example, a glazed fig topped with mascarpone and wrapped with prosciutto is considered an "hors d'oeuvre", whereas figs on a platter are not.

Examples of hors d'oeuvres include:

- Canapés
- Caviar
- Cold cuts
- Deviled eggs
- Dumplings
- Bruschetta
- Pigs in a blanket
- Tongue toast
- Spanakopita

Question 1: What is hors d'oeuvre?

Question 2: When will hors d'oeuvre be served during a meal?

Task 2 Further reading.

History of Sandwich

The first written usage of the English word "sandwish" appeared in Edward Gibbon's journal, in longhand, referring to "bits of cold meat" as a "Sandwich". The *Oxford English Dictionary* gives its appearance in 1762. It was named after John Montagu, the 4th Earl of Sandwich, an 18th-century English aristocrat, although he was neither the inventor nor sustainer of the food. It is said that he ordered his valet to bring him meat tucked between two pieces of bread, and because Montagu also happened to be the 4th Earl of Sandwich, others began to order "the same as Sandwich"! It is said that Lord Sandwich was fond of this form of food because it allowed him to continue playing cards, particularly cribbage while eating without getting his cards greasy from eating meat with his bare hands.

Bread has been eaten together with some meat or vegetables since the Neolithic Age. For example, the ancient Jewish sage Hillel the Elder is said to have placed meat from the Korban Pesach lamb and maror herbs between two pieces of matzah (or flat, unleavened bread) during Passover. Thick slabs of coarse and usually stale bread, called "trenchers", were used as plates. After a meal, the food-soaked trencher was fed to a dog or to beggars, or eaten by the diners. Trenchers were the precursors of Open sandwich |open-face sandwiches.

The immediate cultural precursor with a direct connection to the English sandwich was to be found in the Netherlands of the 17th century, where the naturalist John Ray observed topographical, moral, & physiological; made in a journey through part of the Low Countries, Germany, Italy, and France… (vol. I, 1673) quoted in Simon Schama, *The Embarrassment of Riches* (1987: 152). that in the taverns beef hung from the rafters "which they cut into thin slices and eat with bread and butter laying the slices upon the butter"—explanatory specifications that reveal the Dutch "belegde broodje" was as yet unfamiliar in England.

Initially perceived as food the lower-class people shared while gaming and drinking at night, the sandwich gradually became popular in polite society as it was often served at a late-night meal among the aristocracy. The sandwich's popularity in Spain and England increased dramatically during the 19th century, when the fast pace of an industrial society and the continuously enlarged working classes made fast, portable and inexpensive meals essential.

It was at the same time that the sandwich finally began to appear beyond Europe. In the United States, the sandwich was first promoted as an elaborate meal at supper. By the early 20th century, as bread became a staple food of the United States diet, the sandwich became the same kind of popular, quick meal in the Mediterranean.

Unit 2 Soups and Salads

Learning goals

To know essential information of various kinds of soups and salads, including thick soup, puree soup and so on.

To get to know the making process of them.

Vocabulary

broth [brɒθ] *n.* 肉汤;清汤
salad [ˈsæləd] *n.* 色拉;凉拌生菜;杂烩
dressing [ˈdresɪŋ] *n.* 调料
thick soup 浓汤
puree soup 茸汤
consommé German 德式清炖肉汤
cream soup 奶油稀汤
bisque [bɪsk] *n.* 浓汤
clear soup 清汤
iced soup 冷汤
French onion soup 法式洋葱汤
vichyssoise [ˌviːʃiːˈswɑːz] *n.* 奶油浓汤
chowder [tʃaʊdə(r)] *n.* 〈美〉杂烩(一种用鲜鱼、咸肉与洋葱等煨成的食品)
caprese [kæpˈriːz] *n.* 红白小碟
fattoush 阿拉伯蔬菜沙拉
coleslaw [ˈkəulslɔː] *n.* 凉拌卷心菜
gravy [ˈgreɪvi] *n.* 肉汁;肉汤
sauce [sɔːs] *n.* 酱汁;调味汁
waldorf salad 沃尔多夫色拉
tabbouleh [təˈbuːleɪ] *n.* 塔博勒色拉(一种黎巴嫩生菜)
rustic Greek salad 希腊乡村沙拉

duck salad with plum dressing 李子酱鸭沙拉
seared scallop and mango salsa salad 烤扇贝和芒果沙拉
warm chicken and mushroom salad 热鸡肉和蘑菇沙拉
warm fennel and cherry tomato salad 热茴香和樱桃番茄沙拉

Dialogue

(*C1=Commis*, *C2=Chef*.)

C1: What should I do?

C2: Put the meat into a large saucepan with the butter, onion, parsley, paprika, saffron and salt.

C1: And after that?

C2: Cook gently while stirring gently for 5 minutes.

C1: I've finished.

C2: Good. Then cover the pan and cook it at a simmer.

C1: For how long?

C2: For one and a half hours.

C1: Add the rice?

C2: Yes, and cook for 15 to 20 minutes.

C1: Can I drop the egg into the soup quickly?

C2: No. Let the egg drop slowly while constantly stirring the soup.

C1: The soup is done.

C2: Good. Pour the soup into a heated tureen and serve it immediately.

Activity 1

Task Try to translate the names of those dishes into Chinese.

1. crab Louie salad

2. green salad

3. potato salad with egg and mayo

4. pasta salad closeup
5. tuna salad
6. green papaya salad
7. caesar salad
8. chef salad
9. lobster soup
10. vegetarian minestrone soup
11. cream pumpkin soup
12. French onion soup

Activity 2

Task 1 French Bean Salad.

Ingredients:

French bean

1 sweet onion

Several beans

Mustard

Vinegar

Olive oil

Directions:

Combine the mustard and vinegar in a mixing bowl.

Emulsify with olive oil.

Add the chopped onion and beans then season to taste.

Task 2 Smoked eel with French Bean Salad and Carrot Puree.

Ingredients:

Smoked eel fillet, without skin

100 g French beans, blanched and sliced

20 g shallots, chopped

2 ml white wine vinegar

2 g Dijon mustard

20 ml extra virgin olive oil

100 ml carrot juice

100 g confit carrots

20 ml extra virgin olive oil

Directions:

Combine the mustard and vinegar in a mixing bowl. Emulsify with olive oil. Add the chopped shallots and beans, then season to taste.

To make the carrot puree, pour the carrot juice into a saucepan and reduce by half. Blend the carrot juice and confit carrot. Pass through a sieve and finish with olive oil. Season to taste.

Garnish with horseradish cream, caviar and a sprig of chervil.

Task 3 Seafood Pasta Salad.

Ingredients:

250 g short pasta

1 red pepper

1 cucumber

1 small onion

Half cup of olives

2 hardboiled eggs

Half cup of mayonnaise

1 can of tuna

Any kind of seafood, like prawns, mussels, etc.

Salt and pepper

Directions:

First, boil the pasta and let it cool down.

Then chop up all the ingredients.

Put everything in a bowl and add the mayonnaise, mix and serve hot.

Task 4 French Onion Soup.

Ingredients:

3 tablespoons of unsalted butter

1 small red onion, thinly sliced

3 white onions, thinly sliced

1 clove of garlic, finely chopped

3 tablespoons of all-purpose flour

1 cup of white wine

6 cups of brown stock or water

1 bouquet garni

1 tablespoon of sherry

12 French baguette slices

2 cups of finely grated cheese

Directions:

Melt the butter in a large heavy-bottomed saucepan over medium heat. Add the onions and cook for about 20 minutes, stirring often, until caramelized and dark golden brown. Pour the garlic and the flour into the saucepan and cook, stirring constantly, for 1 to 2 minutes.

Add the white wine and stir the mixture until the flour has blended in smoothly. Bring to a boil slowly, stirring constantly. Whisk or briskly stir in the stock or water, add the bouquet garni and season to taste with salt and freshly ground black pepper. Simmer gently for 30 minutes, then skim the surface of any excess fat if necessary. Add the sherry to the soup and adjust the seasoning to taste.

To make the croutes, toast the bread slices under the broiler until dry and golden on both sides.

Ladle the soup into warm flameproof bowls and float a few croutes on top of each. Sprinkle the top of each soup with Gruyere cheese and place the bowls under a preheated broiler until the cheese melts and becomes golden brown. Serve the soup immediately.

Task 5 Chicken Caesar Salad.

Ingredients:

For the salad:

4 slices of prosciutto

2 cornfed chicken breasts, bone and skin removed

1 knob of butter

2 thick slices of white bread, cubed

2 cos lettuces

For the dressing:

2 garlic cloves

75 ml/3 fl oz white wine

2 eggs, yolks only

1 anchovy fillet

75 g/3 oz parmesan cheese, grated

150 ml/5 fl oz extra virgin olive oil

2 teaspoons of Dijon mustard

Salt and pepper

Directions:

Preheat the grill to high. Place the prosciutto slices onto a baking sheet and place under the grill for 3 to 4 minutes, or until crisp. Remove the prosciutto slices and place on a plate lined with kitchen paper to drain.

Cut both chicken breasts through the middle, without cutting through, to form one large flat piece. Season, to taste, with salt and black pepper and brush with olive oil.

Heat a griddle pan until hot then place the chicken breasts on it and cook for 2—3 minutes on each side, or until just cooked through.

Heat a frying pan until hot and add the butter and then the cubed bread. Fry until golden brown, then remove.

Separate the leaves from the lettuce and cut into chunky pieces.

For the dressing, in a medium saucepan, bring the garlic and wine to a boil and simmer for about 5 minutes, until the garlic has been softened. Leave to cool.

Combine the wine and garlic with the egg yolks, anchovy and cheese in a mixing bowl. Blend with a hand blender or food processor until smooth. Drizzle in the oil in a thin steady stream, taking care not to add it too quickly, otherwise it could split and curdle. Stir in the mustard and add seasoning to taste.

Add the crisp-fried prosciutto, croutons and lettuce to the dressing and toss to combine. To serve, place salad on a plate, top with chicken and pour the dressing over.

Activity 3

Task 1 Reading.

Salad

Salad is a category of dish whose basic ingredient is usually raw vegetables. Salad has at least three ingredients often served with a sauce or dressing including oil and an acid as a light savory dish. The category also includes a variety of related dishes, such as cold cooked vegetables with grains and pasta, add cold meat or seafood, sweet dishes made of cut-up fruit; and even warm dishes. Though the prototypical salad is light, a dinner salad can constitute a complete meal.

Green salads include leaf lettuce and leafy vegetables with sauces or dressings. Although most salads are served cold, some, such as south German potato salad, are served warm.

Salads are generally served with a dressing, as well as various garnishes such as nuts or croutons, and sometimes with the addition of meat, fish, pasta, cheese, eggs, or whole grains.

Salads may be served at any point during a meal. Appetizer salads, light salads stimulate the appetite as the first course of the meal. Side salads accompany the main course as a side dish. Main course salads, usually contain a portion of protein, such as chicken breast or slices of beef. Palate-cleansing salads settle the stomach after the main course. Dessert salads usually contain fruit, gelatin or whipped cream.

Question 1: What is salad?

Question 2: How many kinds of salads do you know according to the function?

Soup

Soup is a primarily liquid food, generally served warm (but may be cool or cold), combining ingredients such as meat and vegetables with stock, juice, water, or another liquid. Hot soups are additionally characterized by boiling solid ingredients in liquids in a pot until the flavors are extracted, forming a broth.

Traditionally, soups are classified into two main groups: clear soups and thick soups. The established French classifications of clear soups are bouillon and consommé. Thick soups mainly include the following types: purées are vegetable soups thickened with starch; bisques are made from puréed shellfish or vegetables thickened with cream; cream soups may be thickened with béchamel sauce; veloutés are thickened with eggs, butter, and cream. Other ingredients commonly used in thicken soups and broths include rice, lentils, flour, and grains. Many popular soups also include carrots and potatoes.

Soups are similar to stews, and in some cases there may not be a clear distinction between the two; however, soups generally have more liquid than stews.

Question 1: What is soup?

Question 2: What different between clear soup and thick soup?

Task 2 Further reading.

Global and Regional Cuisines

Global cuisines can be categorized by various regions according to the common use of major foodstuffs, including grains, produce and cooking fats. Regional cuisines may vary based upon food availability and trade, cooking traditions and practices, and cultural differences. For example, in Central and South America, corn (maize), both fresh and dried, is a staple food. In northern Europe, wheat, rye, and fats of animal origin predominate, while in southern Europe olive oil is ubiquitous and rice is more prevalent. In northern Italy the cuisine featuring butter and rice, stands in contrast to that in southern Italy, featuring wheat pasta and olive oil. China likewise can be divided into rice regions and noodle and bread regions. Throughout the Middle East and the Mediterranean there is a common thread marking the use of lamb, olive oil, lemons, peppers, and rice. The vegetarianism practiced in much of India has made pulses (crops harvested solely for the dry seed) such as chickpeas and lentils as significant as wheat or rice. From India to Indonesia the use of spices is characteristic; coconuts and seafood are used throughout the region both as foodstuffs and as seasonings.

Types of Salads

Green salad

The "green salad" or "garden salad" is most often composed of leafy vegetables such as lettuce varieties, spinach, or rocket (arugula). Due to their low caloric density, green salads are a common diet food. The salad leaves may be cut or torn into bite-sized fragments and tossed together (called a tossed salad), or may be placed in a predetermined arrangement (a composed salad).

Vegetable salad

Vegetables other than greens may be used in a salad. Common raw vegetables used in a salad include cucumbers, peppers, tomatoes, mushrooms, onions, spring onions, red onions, carrots, celery, and radishes. Other ingredients, such as avocado, olives, hard boiled eggs, artichoke hearts, heart of palm, roasted red bell peppers, green beans, croutons, cheeses, meat (e.g. bacon, chicken), or seafood (e.g. tuna, shrimp), are sometimes added to salads.

Bound salad

American-style potato salad with egg and mayonnaise. A "bound" salad can be composed (arranged) or tossed (put in a bowl and mixed with a thick dressing). They are assembled with thick sauces such as mayonnaise. One portion of a true bound salad will hold its shape when placed on a plate with an ice-cream scoop. There is a variety bound salad, such as tuna salad, pasta salad, chicken salad, egg salad, and potato salad.

Bound salads are often used as sandwich fillings. They are also popular at picnics and barbecues, because they can be made ahead of time and refrigerated.

Main course salads

Main course salads (also known as "dinner salads" and commonly known as "entrée salads" in North America) may contain grilled or fried chicken pieces, seafood (grilled or fried shrimp) or a fish steak (tuna, mahi-mahi, or salmon). Sliced steak, such as sirloin or skirt, can be placed upon the salad. Caesar salad, Chef salad, Cobb salad, Greek salad, and Michigan salad are types of dinner salad.

Fruit salads

Fruit salads are made of fruit, and include the fruit cocktail that can be made fresh or from canned fruit.

Dessert salads

Dessert salads rarely include leafy greens and are often sweet. Common variants are made with gelatin or whipped cream, for example, jello salad, pistachio salad, and ambrosia. Other forms of dessert salads include snickers salad, glorified rice, and cookie salad popular in parts of the Midwestern United States.

Unit 3 Eggs and Cheese

Learning goals

To know essential information of various kinds of dishes made from eggs and cheese.

To get to know the making process of some classic dishes.

Vocabulary

cheese [tʃiːz] *n.* 干酪;乳酪;奶酪
cream [kriːm] *n.* 奶油;乳酪
sour butter 酸黄油
yogurt [ˈjɒgət] *n.* 酸奶;酵母乳
pudding [ˈpʊdɪŋ] *n.* 布丁
scrambled egg 炒蛋
egg benedict 烟肉及鸡蛋松饼
frittata 菜肉馅煎蛋饼
eggs Florentine 佛罗伦萨式蛋羹
quiche Lorraine 法式蛋塔
omelet 蛋卷
blue cheese 蓝芝士
cream cheese 奶油奶酪
goat cheese 山羊乳干酪
whey cheese 乳清干酪
smoked cheese 熏制的奶酪

Dialogue

Directions: Practice the following dialogue with your partner.
(C1=*Commis*, C2=*Chef*.)
C1: What kind of omelet should I fix?

C2: A cheese omelet. Have you broken three eggs?

C1: No, not yet.

C2: Break three eggs, please. Mix the eggs with salt and pepper.

C1: I will heat the butter in the omelet pan.

C2: Right.

C1: The butter is browning.

C2: Pour in the eggs... and stir quickly!

C1: Should I use a wooden spoon?

C2: No, use a fork.

C1: What is next?

C2: Sprinkle the omelet with cheese.

C1: OK, now I will fold the omelet.

C2: Right. Now slide the omelet into the plate.

Activity 1

Task Try to translate the names of those dishes into Chinese.

1. cheesecake

2. poutine

3. raclette

4. mozzarella

5. Swiss fondue

Activity 2

Task 1 Poached eggs (Sous Vide cooking).

Cooked at temperature.

Ingredients:

Large hen eggs, duck eggs or quail eggs (quantity is variable)

Directions:

Cooking time: 60 minutes

Set the rear pump flow switch to fully closed. Set the front flow switch to the minimum flow to ensure the delicate proteins in the whites do not separate from agitation.

Set the Sous Vide Professional to the desired temperature based on desired doneness of egg: 143.5°F(62℃)for soft whites, 145.5°F (63℃) for medium set whites or 147.0°F (64℃) for firm set whites.

Once the target temperature is reached, gently place the eggs in the circulating water bath. You may want to use a ladle or slotted spoon to gently lower the eggs so they do not crack.

Cook to the desired doneness for 45 minutes. It's a general rule that most chicken and duck eggs will set to the desired doneness in 60 minutes (approximately 1 minute/gm of egg). Quail eggs will generally cook to the desired doneness in 20—30 minutes.

The proteins will start to denature after 120 minutes, resulting in unpleasant textures.

If plating immediately, gently crack an egg onto a paper towel to capture any excess liquid. Gently roll an egg off of the towel onto a plate.

If serving at a later point, plunge the egg into an ice bath. Store up to 48 hours under refrigeration. Reheat the egg by placing in a 140°F(60℃) circulating bath or placing the cracked egg into a pot of simmering water for 60 seconds.

Task 2 Gougeres.

Preparation time: 25 minutes

Total cooking time: 25 minutes

Amount: 25—30 servings

Ingredients:

1 cream puff pastry

1/3 cup of finely shredded Gruyere or Cheddar cheese

1 egg, beaten

Directions:

Preheat the oven to 32.5°F. Lightly grease two baking sheets. Mix half the cheese into the pastry dough.

Spoon the mixture into a pastry bag fitted with a small plain tip. Pipe out 1-inch balls of dough onto the prepared sheets, leaving a space of 1/4 inches between each ball. Using a fork dipped in the beaten egg, slightly flatten the top of each ball. Sprinkle with the remaining shredded cheese. Bake the balls for 20—25 minutes, or until they have puffed up and are golden brown. Serve hot.

CHEF'S TIP: This is a very simple and light finger food to serve with pre-dinner drinks. Gougeres are sometimes served in restaurants with drinks and referred to as amuse-gueule, the French term for an appetizer.

Activity 3

Task 1 Reading.

Egg Salad

Egg salad is part of a tradition of salads involving a high-protein and low-carbohydrate ingredient mixed with seasonings in the form of spices, herbs, and other foods, and bound with mayonnaise. Its siblings include tuna salad, chicken salad, ham salad, lobster salad, and crab salad.

Egg salad is often used as a sandwich filling, typically made of chopped hard-boiled eggs, mayonnaise, mustard, minced celery, onion, salt, pepper and paprika. It is also often used as a topping on green salads.

A closely related sandwich filler is egg mayonnaise, where chopped hard-boiled egg is mixed with mayonnaise only.

Egg salad can be made creatively with any number of other cold foods added. Onions, lettuce, pickles, pickle relish, capers, bacon, peppers, cheese, celery and cucumber are common ingredients.

In Britain and Ireland, egg salad refers to green or mixed vegetable salad with egg on the side as the protein part of the meal, but not necessarily in mayonnaise.

Question 1: What is egg salad made of?

Question 2: What is egg salad in Britain and Ireland?

Cheese

Cheese is a generic term for a diverse group of milk-based food products. Cheese is produced in wide-ranging flavors, textures, and forms.

Cheese consists of proteins and fat from milk, usually the milk of cows, buffalos, goats, or sheep. It is produced by coagulation of the milk protein casein. Typically, the milk is acidified and the addition of the enzyme rennet causes coagulation. The solids are separated and pressed into final form. Some cheeses have molds on the rind or throughout. Most cheeses melt at cooking temperature.

Hundreds of types of cheese from various countries are produced. Their styles, textures and flavors depend on the origin of the milk (including the animal's diet), whether they have been pasteurized, the butterfat content, the bacteria and mold, the processing, and aging. Herbs, spices, or wood smoke may be used as flavoring agents. The yellow to red color of many cheeses, such as Red Leicester, is normally formed from adding annatto.

For a few cheeses, the milk is curdled by adding acids such as vinegar or lemon juice. Most cheeses are acidified to a lesser degree by bacteria, which turn milk sugar into lactic acid, then the addition of rennet completes the curdling. Vegetarian alternatives to rennet are available; most are produced by fermentation of the fungus Mucor miehei, but others have been extracted from various species of the Cynara thistle family.

Cheese is valued for its portability, long shelf life, and high content of fat, protein, calcium, and phosphorus. Cheese is more compact and has a longer shelf life than milk, although how long a cheese will keep may depend on the type of cheese; labels on packets of cheese often claim that a cheese should be consumed within three to five days after the package is opened. Generally speaking, hard cheeses last longer than soft cheeses, such as Brie or goat's milk cheese. Cheesemakers near a dairy region may benefit from fresher, lower-priced milk, and lower shipping costs. The long storage life of some cheese, especially if it is encased in a protective rind, allows selling when markets are favorable. Additional ingredients may be added to some cheeses, such as black peppers, garlic, chives or cranberries.

A specialist seller of cheese is sometimes known as a cheesemonger. To become an expert in this field, requires some formal education and years of tasting and hands-on experience. This position is typically responsible for all aspects of the cheese inventory; selecting the cheese menu, purchasing, receiving, storage, and ripening.

Question 1: What is cheese?

Question 2: What is the feature of cheese?

Task 2 Further reading.

Types of Cheese

There are many types of cheese, with around 500 different varieties recognised by the International Dairy Federation, over 400 identified by Walter and Hargrove, over 500 by Burkhalter, and over 1,000 by Sandine and Elliker. The varieties may be grouped or classified into types according to criteria such as length of ageing, texture, methods of making, fat content, animal milk, country or region of origin, etc. —with these criteria either being used singly or in combination, but with no single method being universally used. The method most commonly and traditionally used is based on moisture content, which is then further discriminated by fat content and curing or ripening methods. Some attempts have been made to rationalise the classification of cheese: Pieter Walstra proposed a scheme, which uses the primary and the secondary starters combined with moisture content; Walter and Hargrove suggested classifying by production methods which produces 18 types, which are then further grouped by moisture content.

Moisture content (soft to hard)

Categorizing cheeses by firmness is a common but inexact practice. The lines between "soft" "semi-soft" "semi-hard" and "hard" are arbitrary, and many types of cheese are made in softer or firmer variations. The main factor that controls cheese hardness is moisture content, which depends largely on the pressure with which it is packed into molds, and on aging time.

Fresh, whey and stretched curd cheeses. The main factor in the categorization of these cheese is their age. Fresh cheeses without additional preservatives can spoil in a matter of days.

Other content (double cream, goat, ewe and water buffalo)

Some cheeses are categorized by the source of the milk used to produce them or by the added fat content of the milk from which they are produced. While most of the world's commercially available cheese is made from cows' milk, many parts of the world also produce cheese from milk of goats and sheep. Double cream cheeses are soft cheeses of cows' milk enriched with cream so that their fat content is 60% or, in the case of triple creams, 75%.

Soft-ripened, washed rind and blue-vein

There are at least three kinds of cheese according to the presence of mold: soft ripened cheeses, washed rind cheeses and blue cheeses.

Processed cheeses

Processed cheese is made from traditional cheese and emulsifying salts, often with the addition of milk, more salt, preservatives, and food coloring. It is inexpensive and consistent, and melts smoothly. It is sold packaged and either pre-sliced or unsliced, in a number of varieties. It is also available in aerosol cans in some countries.

Eating and cooking cheese

Zigerbrüt, cheese grated onto bread through a mill, from the Canton of Glarus in Switzerland.

At refrigerator temperatures, the fat in a piece of cheese is as hard as unsoftened butter, and its protein structure is stiff as well. Flavor and odor compounds are less easily liberated when cold. For improvements in flavor and texture, it is widely advised that cheeses be allowed to warm up to room temperature before eating. If the cheese is further warmed, to 26—32℃ (79—90°F), the fats will begin to "sweat out" as they go beyond soft to fully liquid.

Above room temperatures, most hard cheeses melt. Rennet-curdled cheeses have a gel-like protein matrix that is broken down by heat. When enough protein bonds are broken, the cheese itself turns from a solid food to a viscous liquid. Soft, high-moisture cheeses will melt at around 55℃ (131°F), while hard, low-moisture cheeses such as Parmesan remain solid until they reach about 82℃ (180°F). Acid-set cheeses, including halloumi, paneer, some whey cheeses and many varieties of fresh goat cheese, have a protein structure that remains intact at high temperatures. When cooked, these cheeses just get firmer as water evaporates.

Some cheeses, like raclette, melt smoothly; many tend to become stringy or suffer from a separation of their fats. Many of these can be coaxed into melting smoothly in the presence of acids or starch. Fondue, with wine providing the acidity, is a good example of a smoothly melted cheese dish. Elastic stringiness is a quality that is sometimes enjoyed, in dishes including pizza and Welsh rarebit. Even a melted cheese eventually turns solid again, after enough moisture is cooked off. The saying "you can't melt cheese twice" (meaning "some things can only be done once") refers to the fact that the oil of the cheese leaches out during the first melting and are gone, leaving the non-meltable solids behind.

As its temperature continues to rise, cheese will brown and eventually burn. Browned, partially burned cheese has a particular distinct flavor of its own and is frequently used in cooking (e.g, sprinkling atop items before baking them).

Unit 4　Fish and Seafood

Learning goals

To know essential information of various kinds of fish and seafood dishes.

To get to know the making process of some classic dishes.

Vocabulary

gravlax　[ˈgrævlæks] *n.* (用盐、黑胡椒、小茴香、酒等腌制的)渍鲑鱼片
kedgeree　[ˈkedʒəriː] *n.* (印度)鸡蛋葱豆饭
crab roulade 蟹肉卷
ceviche　[səˈviːtʃeɪ] *n.* 酸橘汁腌鱼
crab cake 蟹饼
fritto misto (拌了面粉后炸的)油炸海产(或肉类、蔬菜);油煎杂拌
trout flans 鳟鱼
moules mariniere 烧贻贝
seared tuna 烤金枪鱼
sole meuniere 煎鳎鱼

Dialogue

(C1=*Commis*, C2=*Chef*.)

C1：What should I do?

C2：Cook the onions，please.

C1：I've already done it.

C2：Good. Then please mix the sesame seeds，water，garlic，salt，lemon juice，and red pepper.

C1：How much lemon juice?

C2：20 tablespoons.

C1：What is next?

C2: Sprinkle the baking dish with breadcrumbs and parsley.

C1: Should I put the fish in the baking dish then?

C2: Yes. Pour the sesame seeds and onions over the fish.

C1: Should I cover the fish?

C2: No. Have you lighted the oven?

C1: No.

C2: Light the oven, please. Cook the fish at 400 degrees.

C1: For how long?

C2: For 20 to 25minutes.

C1: After it is cooked, I will garnish the fish with the parsley and olives.

Activity 1

Task Try to translate the names of those dishes into Chinese.

1. bouillabaisse 2. ceviche 3. crab stick

4. fish and chips 5. fish ball 6. fish chowder

7. gefilte fish 8. paella

Activity 2

Task 1 Langoustine and White Bean Panna Cotta.

Ingredients:

200 ml langoustine stock (see basics)

80 g white beans, soaked in water for 24 hours

2 g tarragon

20 ml cream

20 ml olive oil

1 gelatine sheet, soaked

30 g white beans, cooked and peeled

20 g tomato concasse

4 sprigs of baby cress

Prsces lemon dressing

Directions:

Cook the beans in langoustine stock with tarragon until tender. Blend in a food processor until smooth then pass through a sieve. Add the cream and gelatine sheet. Finish with olive oil and season to taste. Pour the panna cotta into glasses and leave to set.

Steam the langoustines for 3 minutes and refresh in ice water. Remove the flesh from the shell then clean and slice. Prepare a salad with the white beans, tomato concasse and baby cress. Adjust seasoning to taste. Present on top of the panna cotta to serve.

Task 2 Tuna Carpaccio with Lobster and Avocado.

Ingredients:

250 g tuna, centre cut

2 avocados, peeled and cubed

150 g lobster meat, cooked and cubed

5 ml olive oil

10 ml lemon juice

80 g tomato concasse

20 g shallots, chopped
2 g basil, chopped
5 ml lemon juice
2 ml olive oil

Directions:

For the tuna Carpaccio, drizzle lemon oil on the tuna and season to taste. Wrap in cling film and freeze until required.

For the lobster and avocado salad, combine the ingredients in a mixing bowl and adjust seasoning. To prepare the tomato salsa, mix all the ingredients together. Again, season to taste.

To assemble, slice the tuna thinly and arrange on a plate. Divide the lobster and avocado salad equally into four round moulds. Top with the tomato salsa. Garnish with boiled quail eggs and caviar. Drizzle each plate with Pisces lemon aioli and balsamic reduction.

Task 3 Honey Baked Cod.

Ingredients:

2 cod slices
Marinade:
6 tablespoons of soy sauce
5 tablespoons of honey
1 tablespoon of Chinese cooking wine
1 tablespoon of water

Directions:

Mix the marinade well.

Pour over the the cod and marinate for 30 minutes.

Preheat oven at 160℃.

Lay your baking tray with foil or Glad baking paper (which is what I use because it doesn't stick to the fish).

Grill the cod for 20 minutes skin-side down before turning over.

Grill the other side for another 12 minutes or till fully cooked.

Serve immediately with salad.

Task 4　Milk Poached Halibut Flavored with Rosemary and Kaffir Lime Leaves (Sous Vide cooking)

Ingredients:

4 portions of halibut, each 6 oz (170 g) and 1—1/2 inch (40 mm) thick

1/2 C(120 ml) milk

3 sprigs rosemary

4 kaffir lime leaves, thinly sliced

Kosher salt, to taste

Directions:

Cooking time: 12—20 minutes

Amount: 4 servings

Set the temperature on your Sous Vide Professional to 125°F(51.7℃).

Season the halibut with salt. Place the halibut in a bag; add milk, rosemary and lime leaves and vacuum seal. Make sure that the fish is not overlapping in the bag.

Drop the bag into the 125°F (51.7℃) water and make sure that it is completely submerged. If necessary, place a small weight on the bag to weight down the fish. After 11 minutes, remove the bag from the water and feel the fish for doneness. If the fish is not done, return the bag back to the water. Check every 2 minutes until the fish has reached desired doneness.

Gently remove halibut from the vacuum bag. If a sear is desired, gently dry off the portion with paper or kitchen towel. Season as desired and sear in a hot pan with olive oil or butter. The halibut may also be grilled, if desired.

Task 5　Salmon (Sous Vide cooking).

Ingredients:

4 center-cut salmon filets with medium fat content, pin bones removed, chilled, 6 oz (170 g) or 3/2 inch (40 mm) thick

1 T (15 ml) extra virgin olive oil

1 bay leaf

Kosher salt and coarse ground black pepper, to taste olive oil or clarified butter for searing

Directions:

Cooking time: 20 minutes

Amount: 4 servings

Step 1: Set the Sous Vide Professional to the desired temperature, with rear pump flow switch set to fully open. For medium rare salmon, 125°F (51.7℃) is found to be the best temperature if it has a medium level fat content.

Step 2: In a small vacuum bag, place next to each other the seasoned, trimmed portions of salmon along with 1/2 T extra virgin olive oil and a half bay leaf.

Step 3: Seal portion to desired vacuum. For delicate fish, the best vacuum percentage is 80%—90%. This will ensure the flesh of the fish portion is not compressed under vacuum, therefore guaranteeing the integrity of the delicate muscle fibers.

Step 4: Once target temperature of 125°F (51.7℃) is reached, place salmon in a circulating water bath.

Step 5: Cook to desired doneness for 12—20 minutes. With salmon, the albumen, or white protein present in the fish, will begin to emerge from the flesh. Once this is barely visible, the fish is ready to remove from the bath.

Step 6: Gently remove fish from the vacuum bag. If a sear is desired, gently dry off the portion with paper or kitchen towel. Season as desired and sear in a hot pan with olive oil or butter.

Activity 3

Task Reading

Sous Vide Cooking

Chefs all over the world have enthusiastically embraced Sous Vide cooking which relies on precise temperature control to achieve amazing flavor and texture. Food is vacuum-sealed and cooked at a gentle temperature in a precisely controlled water bath for perfect, repeatable results every time.

Although the term "Sous Vide" could be translated into "under vacuum", a better name for the technique would "be precise temperature cooking". Food is packaged with a vacuum sealer in heat and food-grade plastic pouches before being cooked at a gentle temperature in a precisely controlled water bath. Removing air from the bag guarantees that the food is fully submerged in the water and evenly cooked from all sides.

Sous Vide cooking is ideal for cooking delicate foods like fish and lobsters. It's also great for retaining vibrant flavor and texture in vegetables or to enable lengthier cook tines on secondary cuts of meat without drying them out.

This new dimension in temperature control for your kitchen is easy to learn and presents a completely different cooking experience.

Benefits of Sous Vide cooking

Culinary:

Exact doneness for delicate foods

Moist and tender texture

Enhanced flavors

Perfect results that are easy to repeat

Retention of nutrients

Ability to maintain food at serving temperature for an extended time without overcooking.

Economic:

Less shrinkage and up to 30% more yield

Secondary cuts of meats as tender as expensive primary cuts

No waste from overcooking

Ability to pre-cook and balance workload

Perfect portion control

Easy to learn and less training required.

Question 1: What is Sous Vide cooking?

Question 2: What is the benefits of Sous Vide cooking?

Unit 5 Poultry and Meat

Learning goals

To know essential information of various kinds of dish made of meat and poultry.

To get to know the making process of them.

Vocabulary

chicken liver pate 鸡肝酱
brochette [brɒˈʃet] *n.* 烤肉叉，小串烤肉
roast chicken 烤鸡
stuffed turkey 火鸡
deviled poussin 烤童子鸡
basque [bɑːsk] *n.* 巴斯克语
beef wellington 威灵顿牛肉馅饼
beef stew 红烩牛肉
lamb rib chop 羊肋排
cassoulet [ˌkæsʊˈleɪ] *n.* 豆焖肉

Dialogue

(*C1=Commis*, *C2=Chef*.)

C1: What should I cook?

C2: Beef stroganoff.

C1: Should I start with the beef?

C2: Yes. Please cut the beef into strips.

C1: I am finished.

C2: Cook the mushrooms, onions, and garlic in butter.

C1: Should I cook them in a skillet?

C2: Yes. Cover and simmer the mushrooms, onions, and garlic for 8 minutes.

C1：What else?

C2：Stir it occasionally. Then remove the vegetables from the skillet.

C1：OK，now I will cover the beef that is in the skillet.

C2：Cook the beef over medium heat.

C1：For how long?

C2：10 minutes.

C1：Time's up.

C2：Add the water，bouillon，salt，and pepper.

C1：I will heat it to a boil.

C2：Then reduce the heat. Cover and simmer the beef.

C1：For how long?

C2：5 to 15 minutes.

C1：The beef is done.

C2：Please add the vegetable mixture of mushrooms，onions，and garlic.

C1：I will heat it to a boil，and then reduce the heat.

C2：Then stir in the sour cream and mustard.

C1：I will garnish the beef stroganoff with some parsley.

C2：Serve it with noodles.

Activity 1

Task 1 Coronation chicken.

Ingredients：

1.5 kg free-range chicken

4 spring onions，sliced

6 garlic cloves，peeled，left whole

1 tablespoon of sea salt

10 whole black peppercorns

For the sauce：

1 tablespoon of vegetable oil

1 onion, chopped

1 tablespoon of curry powder

1 tablespoon of tomato purée

85 ml red wine

150 ml water

1 bay leaf

Salt and freshly ground black pepper

Caster sugar

2 lemon slices

Squeeze of lemon juice

425 ml good-quality mayonnaise

2 tablespoons of apricot purée (made by blending 4 or 5 stoned, dried apricots with 3 tablespoons of water)

3 tablespoons of whipped cream

1 tablespoon of roughly chopped coriander

Rice, cooked according to packet instructions

Directions:

Place the chicken into a saucepan with a tight-fitting lid, cover with water and add the spring onions, garlic, salt and peppercorns.

Bring to the boil and simmer for 30 minutes, turning the chicken once during the cooking process. Cover with a lid and switch off the heat. Leave for 1 hour, then remove the chicken, allow to cool completely, and tear the chicken from the bones into rough pieces.

Meanwhile, for the sauce, heat the oil in a pan over a medium heat. Add the onion and cook gently for 3—4 minutes. Add the curry powder and cook for a further 2 minutes, stirring well. Add the tomato purée, wine, water and bay leaf.

Bring the mixture to the boil. Season with salt and freshly ground black pepper, then add the sugar, lemon slices and lemon juice, to taste. Reduce the heat until the mixture is simmering, uncovered, for 5 — 10 minutes. Strain the sauce through a fine meshed sieve and set aside to cool.

Gradually fold in the mayonnaise and apricot purée, to taste. Add more lemon juice, as necessary.

Fold in the whipped cream. To serve, arrange the portioned chicken on a large platter, pour the sauce over the top, sprinkle with the coriander and serve immediately, with rice.

Task 2 Roast Chicken Tacos.

Ingredients:

1 chicken

1 onion

1 avocado

1 lime

Cilantro

Salt and pepper

Several patties

Directions:

Wash the chicken, season with salt and pepper.

Preheat the oven to 450 degrees.

Roast the chicken at 450 degrees for 35 minutes.

Cut the lemon by half.

Slice avocado into cubes, scoop the cubes out.

Chop the white onion into tots.

Chop the cilantro fine.

Grill the patties until they become golden brown.

Chop the chicken into slices.

Place onion tots, cilantro, avocado, and chicken on the patties.

Squeeze lemon on the top.

Task 3 BBQ Wings.

Ingredients:

10 chicken wings

1 teaspoon of turmeric powder

1 teaspoon of chili powder

1 tablespoon of soy sauce

1 lemon grass-crushed

1 teaspoon of sugar

1 tablespoon of grounded peanuts

1 tablespoon of oil

1 tablespoon of cumin

1 red shallot, chopped

Directions:

Cut the chicken wings at the joints and marinate them with all the ingredients.

Leave for an hour.

Grill chicken wings in a hot oven for about 1/2 hour.

Serve hot.

Task 4 Beef Filets (Sous Vide cooking).

Ingredients:

4 beef filets, about 2 inches (50 mm) thick

56 g unsalted butter

2 shallots, each cut in half

4 sprigs of thyme

Kosher salt and coarse-ground black pepper, to taste olive oil or butter for searing

Directions:

For medium rare, set the Sous Vide Professional to the 135°F (57.2 ℃), with rear pump flow switch closed and front flow switch set to full open.

Season beef filets with kosher salt and coarse ground black pepper. In 4 small vacuum bags, place 4 seasoned, trimmed beef filets with 1 T (14g) unsalted butter, half a shallot and thyme sprig respectively.

Seal the beef filets to desired vacuum. With beef, a 90% – 95% vacuum is desirable.

Once the target temperature of 135°F (57.2℃) is reached, place the items in a circulating water bath.

Cook to desired doneness, or about 60 minutes. You can hold at this temperature for up to 90 minutes without effecting quality or texture. The internal temperature should reach 135℉(57.2℃) for medium rare beef.

Remove the beef from the vacuum bags. Dry off with paper towel. Season again, lightly, with kosher salt and coarse ground black pepper.

In a hot pan, grill or plancha, quickly sear off beef filets until browned. This adds additional flavor and texture and is known as Maillard reaction. The optimal result is to have even browning on all sides.

After 60 seconds of rest, the beef is ready to be sliced and plated.

Task 5 Duck Breast (Sous Vide Cooking).

Ingredients:

2 duck breasts, about 1 inch (25 mm) thick

3 T (40 g) duck fat

Kosher salt and black pepper, to taste

Directions:

Set the temperature on your Sous Vide Professional to 135°F (57℃), with rear pump flow switch closed and front flow switch set to full open.

Season the duck breasts with salt and pepper, place them in two bags respectively with duck fat and vacuum seal. Make sure that all the ingredients are cold. If necessary, place the bags with the duck breasts and fat in an ice bath.

Seal them to desired vacuum. For the duck breasts, a vacuum of 90%—95% is optimal. After they are sealed, place the bags on the counter with the skin side down. Shape the duck breasts so the skin side is flat and the duck breasts look plump. This will ensure that you will get even color on the skin side when you render the fat.

Once the target temperature is reached, place the duck into a circulating water bath set to 135°F (57℃) and cook it for 45 minutes.

Remove the bags from the water bath and let them sit on the counter for 10 minutes.

Remove the duck from the bags and place it with the skin side down in a pan on medium heat and render out as much of the fat as you desire. Flip and sear the meat side for no more than 60 seconds. Slice and serve.

Task 6 Chicken Kiev.

Ingredients:

For the chicken kiev:

4 chicken breasts, skin removed

3 garlic cloves, finely chopped

1 lemon, juice only

Salt and freshly ground black pepper

1 tablespoon of chopped fresh tarragon

230 g/8 oz butter, slightly softened

2 tablespoons of plain flour

1 free-range egg, beaten

10—12 tablespoons of fresh breadcrumbs

Olive oil, for frying

For the rice:

250 g/9 oz cooked white and wild rice (from a packet of mixed white and wild rice)

100 g oz bacon lardons, fried until crisp

1 tbsp chopped fresh parsley

100 g/3 oz pine nuts, toasted

2 tablespoons of melted butter

Directions:

Preheat the oven to 200℃/400°F/Gas 6.

For the chicken kiev, slice a piece out of the centre of the chicken breast to make a pocket using a sharp knife. Place the garlic, lemon juice, salt and freshly ground black pepper, tarragon and butter into a bowl and mix them well. Stuff this mixture into the pocket in the chicken breasts.

Dredge the chicken breasts in the flour, then dip them into the beaten egg, then the breadcrumbs to coat completely, shaking off any excess.

Heat the oil in an ovenproof frying pan and fry the chicken breasts on all sides until lightly browned. Transfer them to the oven and bake them for 18—20 minutes, or until golden-brown and completely cooked through.

For the rice, place all the rice ingredients into a bowl and mix well. Pack the rice mixture into four cups or ramekins.

To serve, turn the rice out onto serving plates with the chicken kiev alongside.

Unit 6 Pasta and Rice

Learning goals

To know essential information of various kinds of pasta and rice.

To get to know the making process of them.

Vocabulary

spaghetti [spəˈgeti] *n.* 意大利式细面条
lasagna [ləˈzænjə] *n.* 烤宽面条
linguine [ˈlɪŋgwiːniː] *n.* (意大利)扁面条
ravioli [ˌræviˈəuli] *n.* 意大利式饺子(煮熟后浇番茄酱)
cannelloni [ˌkænəˈləuni] *n.* (意大利的)烤碎肉卷子
gnocchi [ˈnjɒki] *n.* (面粉或马铃薯做的)汤团;团子;通心粉
paella [paɪˈelə] *n.* (西班牙)肉菜饭
risotto [rɪˈsɔːtəʊ] *n.* 意大利调味饭
rice cake 年糕
flour [flaʊə(r)] *n.* 面粉;粉末
semolina [ˌseməˈliːnə] *n.* 粗粒小麦粉
couscous [ˈkuskus] *n.* 蒸粗麦粉
millet [ˈmɪlɪt] *n.* 小米,黍
corn flour 〈美〉玉米粉; 玉米淀粉
cornmeal [ˈkɔːnmiːl] *n.* 玉蜀黍粉;玉米片;麦片
tortellini [ˌtɔːtəˈliːni] *n.* (意大利式)圈形肉馅水饺

penne [ˈpeneɪ] *n.* 短管状通心面
farfalle [fɑːˈfæleɪ] *n.* 蝶形面食;蝴蝶面
rigatoni [ˌrɪgəˈtəʊni] *n.* 波纹贝壳状通心粉
fusilli [fuˈziːli]*n.* 螺旋形意大利面制品
fettucine [ˈfetəsɪn] *n.* 黄油酱汁面条;通心粉的一种
cannelloni [ˌkænəˈləuni] *n.* (意大利)烤碎肉卷子
ziti [ˈziːti] *n.* 意大利通心面
macaroni [ˌmækəˈrəuni] *n.* 通心粉
spinach tagliatelle 菠菜意大利面

Dialogue

A: Have you ever tried Italian food?
B: Yes, several times already.
A: What did you eat?
B: Spaghetti.
A: How do you like it?
B: Well, you know I am used to eating Chinese food which is spicy.
A: Oh, exactly. I was puzzled to see some Chinese put extra spice onto Italian noodles. You know it can spoil the food.
B: But I understand them now.

Activity 1

Task 1 Creamy Chicken, Bacon and Basil Pasta.

Ingredients:
3 tablespoons of olive oil
3 large boneless chicken thigh, skin removed, cut into strips
5 rashers bacon, chopped
1—2 garlic cloves, crushed
Salt and freshly ground black pepper
300 ml/10 fl oz double cream
450 g/1 lb farfalle

A handful of fresh basil, torn, plus extra for garnish

200 g/7 oz cheddar or parmesan, grated, plus extra to garnish

Directions:

Cook the pasta according to packet instructions in a pan of salted boiling water, then drain.

Heat the olive oil in a frying pan, add the chicken strips and bacon and cook for 3—4 minutes, or until the chicken is golden-brown and cooked through.

Add the garlic and cook for 1 minute. Season, to taste, with salt and freshly ground black pepper, add the cream and warm through.

Add the creamy sauce to the cooked, drained pasta and stir well.

To serve, stir in the basil and cheddar spoon onto serving plates. Garnish with extra grated cheese and basil leaves.

Task 2 Spaghetti Bolognese.

Ingredients:

2 tablespoons of olive oil or sun-dried tomato oil from the jar

6 rashers of smoked streaky bacon, chopped

2 large onions, chopped

3 garlic cloves, crushed

1 kg lean minced beef

2 large glasses of red wine

2×400 g cans chopped tomatoes

1×290 g jar antipasti marinated mushrooms, drained

2 fresh or dried bay leaves

1 teaspoon of dried oregano or a small handful of fresh leaves, chopped

1 teaspoon of dried thyme or a small handful of fresh leaves, chopped

Drizzle balsamic vinegar

12—14 sun-dried tomato halves, in oil

Salt and freshly ground black pepper

A good handful of fresh basil leaves, torn into small pieces

800 g—1000 g dried spaghetti

Lots of freshly grated parmesan, to serve

Directions:

Heat the oil in a large, heavy-based saucepan and fry the bacon until golden over a medium heat. Add the onions and garlic, frying until softened. Increase the heat and add the minced beef. Fry it until it has browned, breaking down any chunks of meat with a wooden spoon. Pour in the wine and boil until it has reduced in volume by about a third. Reduce the temperature and stir in the tomatoes, drained mushrooms, bay leaves, oregano, thyme and balsamic vinegar.

Either blitz the sun-dried tomatoes in a small blender with a little of the oil to loosen, or just finely chop before adding to the pan. Season well with salt and pepper. Cover with a lid and simmer the Bolognese sauce over a gentle heat for 1—1½ hours until it's rich and thickened, stirring occasionally. At the end of the cooking time, stir in the basil and add any extra seasoning if necessary.

Remove from the heat to "settle" while you cook the spaghetti in plenty of boiling salted water (for the time stated on the packet). Drain and divide between warmed plates. Scatter a little parmesan over the spaghetti before adding a good ladleful of the Bolognese sauce, finishing with a scattering of more cheese and a twist of black pepper.

Task 3 Penne with Spicy Tomato and Mozzarella Sauce.

Ingredients:

2 teaspoons of olive oil

1 onion, peeled, finely chopped

3 garlic cloves, peeled, finely chopped

1 red chilli, seeds removed, finely chopped

1×400 g/14 oz can chopped tomatoes

Splash red wine vinegar

1 tablespoon of tomato purée

Tabasco sauce, to taste

1 tablespoon of sugar

A small bunch of fresh basil leaves, torn

Salt and freshly ground black pepper

1×125 g/4½ oz ball fresh buffalo mozzarella, crumbled

400 g/14 oz dried penne pasta, cooked according to packet instructions, drained

Grated parmesan, to serve

Directions:

Heat the oil in a saucepan over a medium heat. Add the onion and fry for 4—5 minutes, or until softened and beginning to color.

Add the garlic and chilli and continue to fry for 1—2 minutes.

Pour in the chopped tomatoes, red wine vinegar, tomato purée, a splash or two of Tabasco (to taste), sugar and half of the basil leaves, and stir well. Season, to taste, with salt and freshly ground black pepper.

Bring the mixture to a simmer, then continue to simmer for 18—20 minutes, or until the sauce has thickened and reduced in volume.

Just before serving, stir the crumbled mozzarella into the spicy tomato sauce. Stir the drained pasta into the sauce.

To serve, divide the pasta and sauce equally among four serving plates. Sprinkle over the remaining torn basil leaves and the grated parmesan.

Task 4 Chicken and Pea Risotto.

Ingredients:

For the risotto:

Olive oil, for frying

2 onions, sliced

400 g/14 oz arborio risotto rice

Splash white wine

1 l/1¾ pints of hot vegetable stock

Salt and freshly ground black pepper

200 g/7 oz fresh peas, blanched

200 g/7 oz asparagus, blanched

400 g/14 oz baby spinach

1 tables poon of chopped fresh basil

75 ml/2½ fl oz double cream

For the chicken:

Oil, for frying

Salt and freshl y ground black pepper

4 chicken breasts, skin on

Chicken or vegetable stock, to cover

Parmesan shavings, to serve

Directions:

Preheat the oven to 200℃/400℉/Gas 6.

For the risotto, heat the oil in a pan and fry the onion until softened. Add the rice and stir well, then add the wine and simmer until reduced completely. Add a good ladleful of the hot stock and stir continuously. When all this has been absorbed, add more stock. Continue adding more stock, stirring continuously, until the rice is cooked.

Season with salt and freshly ground black pepper and stir in the peas, asparagus, spinach, basil and cream.

For the chicken, season the chicken breasts well with salt and freshly ground black pepper. Heat the oil in a frying pan and place the chicken breasts skin-side down. Fry the chicken breasts on both sides until lightly browned.

Place the chicken breasts in an ovenproof dish and pour in enough stock to come one third of the way up the sides of the chicken breasts. Place in the oven for 15 minutes, or until completely cooked through.

To serve, place the risotto into serving bowls. Slice the chicken and place on top of the risotto, then sprinkle with parmesan shavings.

Task 5 Paella.

Ingredients:

170 g/6 oz chorizo, cut into thin slices

110 g/4 oz pancetta, cut into small dice

2 garlic cloves, finely chopped

1 large Spanish onion, finely diced

1 red pepper, diced

1 teaspoon of soft thyme leaves

¼ tsp dried red chilli flakes

570 ml/1 pint calasparra (Spanish short-grain) rice

1 teaspoon of paprika

125 ml/4fl oz dry white wine

1. 2 1/2 pints chicken stock, heated with ¼ teaspoon of saffron strands

8 chicken thighs, each chopped in half and browned

18 small clams, cleaned

110 g/4 oz fresh or frozen peas

4 large tomatoes, deseeded and diced

125 ml/4 fl oz good olive oil

1 head garlic, cloves separated and peeled

12 jumbo raw prawns, in shells

450 g/1 lb squid, cleaned and chopped into bite-sized pieces

5 tablespoons of chopped flat-leaf parsley

Salt and freshly ground black pepper

Directions:

Heat half the olive oil in a paella dish or heavy-based saucepan. Add the chorizo and pancetta and fry until crisp. Add the garlic, onion and pepper and heat until softened. Add the thyme, chilli flakes and calasparra rice, and stir until all the grains of rice are nicely coated and glossy. Now add the paprika and dry white wine and when it is bubbling, pour in the hot chicken stock, add the chicken thighs and cook for 5—10 minutes.

Now place the clams into the dish with the join facing down so that the edges open outwards. Sprinkle in the peas and chopped tomatoes and continue to cook gently for another 10 minutes.

Meanwhile, heat the remaining oil with the garlic cloves in a separate pan and add the prawns. Fry quickly for a minute or two and then add them to the paella. Now do the same with the squid and add them to the paella too.

Scatter the chopped parsley over the paella and serve immediately.

Activity 2

Task Reading.

How to make Lasagne

— a large bowl
— a 13-by-9-inch baking dish
— aluminum foil
— two minced garlic cloves
— a pound of ricotta cheese
— two boxes of frozen spinach, thawed with the liquid squeezed out
— five handfuls of grated Parmesan cheese
— an egg
— two 15-ounce cans of tomato sauce
— a box of no-boil lasagna noodles
— a pound of mozzarella cheese, sliced

Directions:

Step 1: Ricotta mixture. Heat the oven to 350 degrees Fahrenheit (it will take at least 20 minutes to warm up). Mix together the garlic, ricotta, spinach, half of the parmesan, and the egg in the bowl until smooth.

Step 2: Sauce pan revised. Spread a fifth of the tomato sauce on the bottom of the baking dish.

Step 3: Overlap noodle. Overlap a quarter of the noodles in a layer on top of the sauce. It's OK if they don't completely cover the surface area.

Step 4: Spread sauce. Spread a fifth of the tomato sauce on top of the noodles.

Step 5: Dollop mixture. Dollop a third of the ricotta mixture in a few spots over the noodles and flatten the dollops (the ricotta mixture will spread out when it heats up). Lay a quarter of the mozzarella on top.

Step 6: Add mozarella. Repeat the process by laying a quarter of the noodles in the opposite direction than you did before, top with a fifth of the tomato sauce, a third of the ricotta mixture, and a quarter of the mozzarella. Repeat once more: a quarter of the noodles in the opposite direction, a fifth of the tomato sauce, the last third of the ricotta, and a quarter of the mozzarella.

Step 7: Sprinkle parmesan. Cover with the last quarter of the noodles, top with the last fifth of the tomato sauce and last quarter of the mozzarella, and sprinkle with the remaining Parmesan.

Step 8: Aluminum foil oven. Cover the baking dish with aluminum foil and bake until the lasagne is bubbly around the edges, about 35 minutes.

Step 9: Remove foil backing. Remove the foil and bake another 15 minutes, until the top is bubbly and light golden brown. Let the lasagne rest 10 to 15 minutes before slicing, or it will be too runny.

Unit 7 Vegetables

Learning goals

To know essential information of varies kinds of vegetable dishes.

To get to know the making process of the dishes.

Vocabulary

glazed vegetable 釉面的蔬菜

croquette [krəu'ket] *n.* 炸丸子;炸肉饼

gratin ['grætn] *n.* 烘烤菜肴上的脆皮

mornay [ˈmɔːneɪ] *n.* 茅内沙司；奶油蛋黄沙司
mashed potato 土豆泥；马铃薯泥
stuffed tomato 填馅西红柿
tortilla [tɔːˈtiːə] *n.*（墨西哥）玉米粉薄烙饼
baked bean 甜豆
endive [ˈendaɪv] *n.* 菊苣；荷兰莴苣
ratatouille [ˌrætəˈtuːi] *n.* 蔬菜什锦

Dialogue

(*J1=John*, *J2=Jenny.*)

J1: I will make a Nicoise salad.

J2: Good. Start with the vinaigrette dressing, please.

J1: What do I need?

J2: Olive oil, vinegar, salt, basil, dry mustard, and pepper.

J1: OK. I will mix the ingredients and refrigerate them.

J2: Now let us make the salad.

J1: I already cooked the green beans.

J2: Good. Did you refrigerate them?

J1: Yes. I refrigerated the green beans an hour ago.

J2: Excellent. Now tear the lettuce into bite-size pieces. Place the lettuce in the salad bowl.

J1: What should I do with the green beans, tomatoes, and eggs?

J2: Boil the eggs. Cut the hard-boiled eggs into fourths. Cut the tomatoes into sixths. Place the beans, tomatoes, and eggs on the lettuce.

J1: What should I do with the tuna?

J2: Place it in the center of the salad.

J1: What should I do with the olives, anchovies, and parsley?

J2: Garnish the salad with them.

J1: When should I pour the vinaigrette dressing on the Nicoise salad?

J2: Pour the dressing on the salad just before serving.

Activity 1

Task 1　Cumin Butter Carrots (Sous Vide cooking).

Ingredients:

1 bundle of baby carrots

1/2 unsalted butter

1/2 cumin, freshly ground

Kosher salt, to taste

Directions:

Set the temperature on your Sous Vide Professional to 185°F (85℃), with the rear pump flow switch closed and the front flow switch set to fully open.

Peel the carrots and trim the green tops off.

In a small sauce pan, combine the butter, cumin and salt over low heat until the butter melts and a homogeneous mixture takes shape.

Place the carrots and the butter mixture into a medium vacuum bag and vacuum seal to 99.9%, full vacuum.

Once the target temperature of 185°F (85℃) is reached, place the carrots in a circulating water bath. Cook until tender—about 45 minutes.

Remove the bag from the water bath. Take the carrots out to serve.

Task 2　Red & Golden Baby Beets (Sous Vide cooking).

Ingredients:

14—20 red and golden baby beets, tops removed, peeled and halved, separated according to the color

1/2 fresh orange juice

Zest and fresh pressed juice of one lime

6 whole black peppercorns

1 T extra virgin olive oil

1 T chives, finely chopped

Kosher salt, to taste

Directions:

Set the Sous Vide Professional to 185°F (85℃), with the rear pump flow switch closed and the front flow switch set to fully open.

In a separate vacuum bag, place beets along with orange juice, lime juice and zest and black peppercorns. It's important to seal the yellow and red beets separately in order to maintain individual coloring.

Seal beets to full, 99.9% vacuum.

Once the target temperature of 185℉ (85℃) is reached, place the beets in a circulating water bath.

Cook the beets to desired doneness for 45—75 minutes.

Remove the beets from the vacuum bag.

Task 3 Creamy Vegetable Tarts.

Ingredients:

300 g plain flour

15 g baking powder

100 g butter

25 g ground walnuts

5 eggs

1 roasted onion with 4 char-grilled fresh asparagus spears

1 roasted onion with 2 artichoke halves

1 roasted onion with 1—2 char-grilled red peppers

1 roasted onion with 2—3 soaked sundered tomatoes custard sauce

150ml London Gold, 125g cream cheese

15—25 ml balsamic vinegar

Chopped herbs or flavorings

50 g grated parmesan cheese

Directions:

Make up the pastry and line 6—8 individual tart tins. Bake blind, whisk eggs in a bowl with the London Gold and cream cheese.

Season and strain. Add chopped herbs and balsamic vinegar if desired. Arrange fillings in pre-baked fan causes, pour over a little of the sauce, and sprinkle with parmesan. Bake in a low to moderate oven (150℃/gas mark 4) until the custard is set. Sever with salad leaves, fresh herbs and fried balsamic onion slices. For the roasted onion, peel and cut through almost to the base, fan out, brush with oil, roast until golden and tender. Use as a filling for the tarts or tossed Oita balsamic vinegar in the salad.

Task 4 Mashed Potato with Garlic-infused Olive Oil.

Ingredients:

For the basic recipe:

900 g/2 lb potatoes (Désirée or King Edward)

50 g/2 oz butter

For the garlic-infused olive oil:

3 fat garlic cloves, halved lengthways

8 tablespoons of best quality extra virgin olive oil

Salt and freshly milled black pepper

Directions:

First place the garlic and olive oil in a small saucepan over the gentlest heat possible—a heat diffuser is good for this—and leave for 1 hour for the garlic to infuse and become really soft.

Use a potato peeler to pare off the skins as thin as possible and then cut the potatoes into even-sized chunks, not too small. If they are large, quarter them; if they are small, halve them.

Put the potato chunks in a large saucepan, then pour boiling water over them, add 1 dessertspoon of salt, put on a lid and simmer gently until they are absolutely tender—it should take approximately 25 minutes. The way to tell whether they are ready is to pierce them with a skewer in the thickest part; the potato should not be hard in the centre. And you need to be careful here, because if they are slightly underdone you do get lumps.

When the potatoes are cooked, drain them. Cover them with a clean tea cloth to absorb some of the steam for about 5 minutes, then use an electric whisk on a low speed, begin to break them up using half the garlic and oil. As soon as all that is incorporated, add the rest of the garlic and oil and whisk until smooth, seasoning well with salt and freshly milled black pepper.

Task 5 Cauliflower cheese.

Ingredients:

1 medium head cauliflower, broken into large florets

40 g/1½ oz butter

40 g/1½ oz plain flour

400 ml/14 fl oz milk

1 teaspoon of English mustard

100 g/3½ oz mature cheddar cheese, grated

Salt and freshly ground black pepper

Directions:

Preheat the oven to 190℃/375℉.

Wash the cauliflower thoroughly and place in a large saucepan of salted water. Bring to the boil and cook for 3—5 minutes, until the cauliflower is almost tender, but still fairly firm. Tip into a colander and leave to drain.

To make the sauce, melt the butter in a medium, heavy-based pan and stir in the flour. Cook over a gentle heat for 1 minute. Remove the pan from the heat and gradually add the milk, a little at a time, stirring well between each addition. Return the pan to a medium heat and bring the mixture to the boil, stirring constantly. Simmer for 2 minutes, then remove from the heat.

Technique: Thickening a roux to make béchamel sauce

Stir in the mustard and two thirds of the cheese and set aside. Arrange the cauliflower in an ovenproof baking dish. Carefully pour over the sauce, ensuring the cauliflower is completely covered. Scatter over the remaining cheese and bake for 25 — 30 minutes, until the top is golden-brown and bubbling.

Unit 8 Sauces and Desserts

Learning goals

To know essential information of various kinds of sauces and desserts.
To get to know the making process of some classic sauces and desserts.

Vocabulary

cocoa [ˈkəukəu] *n.* 可可粉;可可饮料
icing sugar (制甜食用的)糖粉
yolk [jəuk] *n.* 蛋黄
tart [tɑːt] *n.* 果馅饼
baking soda 小苏打;碳酸氢钠;发酵粉
mixture [ˈmɪkstʃə(r)] *n.* 混合;混合物
fritter [ˈfrɪtə(r)] *n.* 油炸馅饼
honey [ˈhʌni] *n.* 蜂蜜
white sauce 白汁沙司;白调味汁
mornay sauce 茅内沙司
mayonnaise [ˌmeɪəˈneɪz] *n.* 蛋黄酱;美乃滋
sweet and sour sauce 酸甜酱
thousand island dressing 千岛色拉调味汁
fondue [fɑnˈdu, -ˈdju] *n.* 溶化奶油
brandy butter 白兰地黄油,白兰地白脱
pudding [ˈpʊdɪŋ] *n.* 布丁
strudel [ˈstruːdl] *n.* 以果实或干酪为馅而烤成的点心,果馅卷
mousse [muːs] *n.* 冻奶油甜点;慕思
vanilla [vəˈnɪlə] *n.* 香草
roulade [ruːˈlɑːd] *n.* 肉卷
terrine [təˈriːn] *n.* 肉糜
tiramisu [ˌtɪrəmɪˈsʊ] *n.* 提拉米苏(意大利式甜点)

Dialogue

(*C1=Commis, C2=Chef.*)

C1: I will preheat the oven to 350.

C2: Please grease and flour 2 cake pans.

C1: I've already greased and floured the cake pans.

C2: Good. Then put the flour, sugar, cocoa, baking soda, salt, baking powder, water, shortening, eggs, and vanilla in the mixer.

C1: Should I mix the ingredients at low speed?

C2: Yes. For 30 seconds.

C1: And now?

C2: Mix the ingredients at high speed.

C1: For how long?

C2: For 3 minutes.

C1: I will pour the ingredients into baking pans.

C2: OK, put the 2 pans in the oven.

C1: For how long?

C2: 30 minutes.

C1: And in the meantime?

C2: Prepare the cherry filling, please.

C1: The cakes are ready.

C2: Put the cakes on the wire racks to cool.

C1: OK.

C2: The cakes are cooled. I will spread the whipping cream, cherry filling, and frosting.

C2: Garnish the cakes with chocolate curls and maraschino cherries.

C1: Should I refrigerate them?

C2: Yes, please.

Activity 1

Task Try to translate the names of those dishes into Chinese.

1. flourless chocolate cake

2. baked custard

3. strawberry ice cream

4. cookies

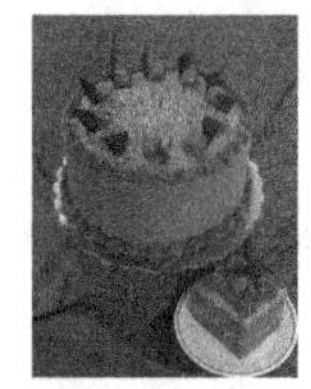
5. chocolate cake

6. pie

7. chocolates

8. cheesecake

9. Czech and Slovak appetizer or snack

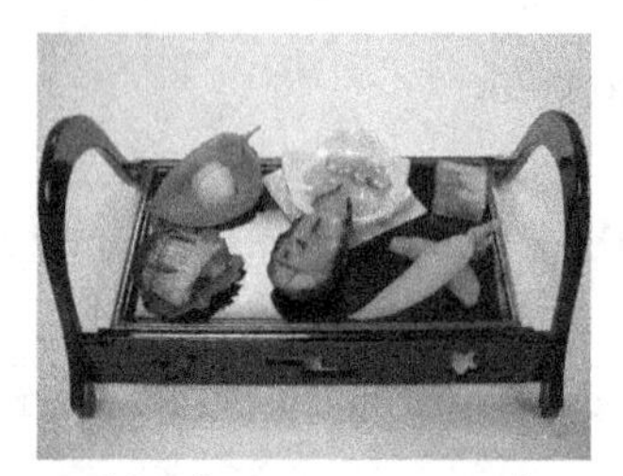
10. Japanese zensai

Activity 2

Task 1 Bread and Butter Pudding.

Ingredients:

25 g/1 oz butter, plus extra for greasing

8 thin bread slices

50 g/2 oz sultanas

2 teaspoons of cinnamon powder

350 ml/12 fl oz whole milk

50 ml/2 fl oz double cream

2 free-range eggs

25 g/1 oz granulated sugar

Nutmeg, grated, to taste

Directions:

Grease a 1 1/2 pint pie dish with butter.

Cut the crusts off the bread. Spread each slice with on one side with butter, then cut into triangles.

Arrange a layer of bread, buttered-side up, in the bottom of the dish, then add a layer of sultanas. Sprinkle with a little cinnamon, then repeat the layers of bread and sultanas, sprinkling with cinnamon, until you have used up all of the bread. Finish with a layer of bread, then set aside.

Gently warm the milk and cream in a pan over a low heat to scalding point. Don't let it boil.

Crack the eggs into a bowl, add three quarters of the sugar and lightly whisk until pale.

Add the warm milk and cream mixture and stir well, then strain the custard into a bowl.

Pour the custard over the prepared bread layers and sprinkle with nutmeg and the remaining sugar and leave to stand for 30 minutes.

Preheat the oven to 180℃/355℉/Gas 4.

Place the dish into the oven and bake for 30—40 minutes, or until the custard has set and the top is golden-brown.

Task 2　Green Apple Split.

Ingredients:

200 ml fresh apple juice

8 g gelatine sheet, soaked in ice water

2 egg whites

200 g sugar

70 ml water

70 g apple puree

10 ml calvados

4 g gelatine sheet, soaked in ice water

35 ml whipped cream

Directions:

To make the apple jelly, bring the apple juice to a simmer. Add the gelatine and allow it to dissolve. Strain the liquid and pour into shot glasses. Refrigerate until set.

Meanwhile, whisk the egg white and add 10gms of sugar to form a peal. Heat remaining sugar with water to a syrup and pour into the egg whites. Dissolve gelatine and whisk with apple puree. Fold the calvados, followed by the egg white and shipped cream.

Task 3　Bourbon Infused Peach (Sous Vide cooking).

Ingredients:

4 peaches, cut in half, pits removed

1 C (240 ml) simple syrup

8 basil leaves, fresh

1/2 C (120 ml) bourbon

Directions:

Set the temperature on your Sous Vide Professional to 180°F (82.2℃), with the rear pump flow switch closed and the front flow switch set to fully open.

In a small saucepan, combine the simple syrup, bourbon and basil leaves.

Bring to a boil. Remove from heat and let sit until cool. Remove the basil leaves.

In a medium vacuum bag, place the peach halves and the bourbon syrup. Vacuum seal to 99.9% vacuum.

Once target temperature of 180° F (82.2℃) is reached, place the peaches in a circulating water bath.

Cook the peaches to desired doneness for 30—45 minutes.

Shock the sealed bag in an ice bath. Open the bag and skin will easily peel from peaches.

Task 4 BBQ Sauce.

Ingredients:

1 small onion, chopped

3 garlic cloves, crushed

Olive oil

1 red chilli, finely chopped

1 teaspoon of fennel seeds, crushed

55 g/2 oz dark brown sugar

50 ml/1¾ fl oz dark soy sauce

300 ml/10 fl oz tomato ketchup

Salt and pepper

Directions:

Fry the onion and the garlic in olive oil with the chilli, fennel seeds and sugar.

Add the soy sauce and ketchup, and season with salt and pepper.

Bring to the boil and simmer for a few minutes to combine the flavors. Use as a dip or to coat spare ribs, chicken or sausages.

Task 5 White Sauce.

Ingredients:

25 g/1 oz butter

25 g/1 oz plain flour

600 ml/1 pint milk

Salt and white pepper

Directions:

Melt the butter in a saucepan.

Stir in the flour and cook for 1—2 minutes.

Take the pan off the heat and gradually stir in the milk to get a smooth sauce. Return to the heat and, stirring continuously till the sauce comes to the boil.

Simmer gently for 8 — 10 minutes and season with salt and white pepper.

Activity 3

Task 1 Reading.

Dessert

Dessert is the usually sweet course that concludes a meal. The food that composes the dessert course includes but is not limited to sweet foods. There is a wide variety of desserts in Western cultures now including cakes, cookies, biscuits, gelatins, pastries, ice creams, pies, pudding, and candies. Fruit is also commonly found in dessert courses because of its natural sweetness. Many different cultures have their own variations of similar desserts around the world, such as in Russia, where many breakfast foods such as blint, oladi, and syrniki can be served with honey and jam to make them popular as desserts. The loosely defined course called dessert can apply to many foods.

Desserts usually contain sugar or a sweetening agent. Desserts contain a range of ingredients which makes the end product differ. Some of the more common ingredients in desserts are flour, dairy, eggs, and spices. Sugar gives many desserts their "addictive sweetness". Sugar also contributes to the moistness of desserts and their tenderness. The flour or starch component in most desserts serves as a protein and gives the dessert structure. Different flours such as all-purpose flour or pastry flour provide a less rigid gluten network and therefore a different texture. Along with flour desserts may contain a dairy product. The extent to which dairy is used is based on the type of dessert. Desserts like ice cream and puddings have some sort of dairy as their main ingredient, whereas desserts like cakes and cookies contain relatively small amounts. The dairy products in baked goods keep the desserts moist. Many desserts also contain eggs, in order to form custard or to aid in the rising and thickening of a cake-like substance. Egg yolks specifically contribute to the richness of desserts. Egg whites can act as a leavening agent when the proteins uncoil and expand. Desserts can contain many different spices and extracts to add a variety of flavors. For example, salt is salty. Salt is added to desserts to balance sweet flavors and create a contrast in flavors. All these ingredients contribute to desserts and make them delicious.

Question 1: What is dessert?

Question 2: When are normal ingredients for making dessert?

Task 2 Further reading.

Sauce

In cooking, a sauce is liquid, creaming or semi-solid food served on or used in preparing other foods. Sauces are not normally consumed by themselves; they add flavor, moisture, and visual appeal to another dish. "Sauce" is a French word taken from the Latin word "salsa", meaning salted. Possibly the oldest sauce recorded is garum, the fish sauce used by the ancient Romans.

Sauces need a liquid component, but some sauces (for example, pico de gallo salsa or chutney) may contain more solid elements than liquid. Sauces is an essential element in cuisines all over the world.

Sauces may be used for savory dishes or for desserts. They can be prepared and served cold, like mayonnaise, prepared cold but served lukewarm like pesto, or can be cooked like bechamel and served warm or again cooked and served cold like apple sauce. Some sauces are industrial inventions like Worcestershire sauce, HP sauce, and nowadays mostly bought ready-made sauces like soy sauce or ketchup. Others still are freshly prepared by the cook. Sauces for salads are called salad dressing. Sauces made by deglazing a pan are called pan sauces. Deglazing is a cooking technique that removes and dissolves browned food residue from a pan to make a sauce.

Varieties of Dissert

Dessert can come in variations of flavors, textures, and looks. Desserts can be defined as a usually sweeter course that concludes a meal. A range of courses anywhere from fruits or dried nuts to multi-ingredient cakes and pies can be called disserts. There are variations of a certain dessert in different cultures. In modern times the variations of desserts have usually been passed down or come from geological regions. This is one cause for the variation of desserts. These are some major categories in which desserts can be placed.

Custards and puddings

These kinds of sweets usually include a thickened dairy base. Custards are cooked and thickened with eggs. Baked custards can include crème brulee and flan. Puddings are thickened with starches.

Frozen desserts

Ice cream and gelato both fit into this category. Ice cream is a cream base that is churned as it is frozen to create a creamy consistency, while gelato uses a milk base and has less air than ice cream. Thirdly, sorbet is made from churned fruit and is not dairy based.

Cakes

Cakes are sweet tender breads made with sugar. Cakes can vary from light, airy sponge cakes to dense cakes with less flour. In addition, small-sized cakes have become popular in the form of cupcakes and petits fours.

Cookies

Cookies are similar to cakes (the word coming from the Dutch word "kowkje" meaning little cake). Historically, cookies were small spoonfuls of cake batter placed in the oven to test the temperature. Cookies can come in many different forms. There are many kinds of cookies, including layered bars, crispy meringues, and soft chocolate chip cookies.

Pies

Pies and cobblers are a crust with a filling. The crust can be made from either a pastry or crumbs. The filling can be anything from fruits to puddings.

Chocolates and candies

Many candies involve the crystallization of sugar which varies in the texture of sugar crystals. Candies can be found in many different forms including caramel, marshmallows, and taffy.

Pastries

Pastries can either take the form of light and flaky bread with an airy texture or unleavened dough with a high fat content. Pastries can be eaten with fruits, chocolates, or other sweeteners.

Miscellaneous desserts

Many desserts cannot be categorized such as the cheesecake. Though a cheesecake is similar to a custard, it is named "cake". Many desserts can span the categories and several don't fit in a category at all.

Unit 9 Baking

Learning goals

To know essential information of various kinds of baking.

To get to know the making process of some classic baked foods.

Vocabulary

baguette [bæˈget] *n.* 法式长棍面包

whole wheat bread 全麦包

sourdough [ˈsaʊə(r)dəʊ] *n.* 酵母

pretzel [ˈpretsl] *n.* 一种脆饼干;椒盐卷饼

roll [rəʊl] *n.* 面包卷

bagel [ˈbeɪgl] *n.* 百吉饼;硬面包圈

challah ['kɑːlə] n. 犹太人在宗教节日所食的鸡蛋面包

bun [bʌn] n. 小面包

stollen ['stəʊlən] n. 〈德〉果子甜面包;葡萄干甜面包

panettone [ˌpænəˈtəʊni] n.(用葡萄干、蜜饯果皮、杏仁等做成的松软的)意大利节日糕点

muffin ['mʌfɪn] n. 松饼

crumpet ['krʌmpɪt] n. 小圆烤饼

pancake ['pænkeɪk] n. 薄煎饼

waffle ['wɒfəl] n. 华夫饼干

pastry ['peɪstri] n. 面粉糕饼;馅饼皮

cookie ['kʊki] n. 饼干

shortbread ['ʃɔːtˌbred] n. 奶油甜酥饼

ginger snaps 姜饼

brownie ['braʊni] 果仁巧克力小方块蛋糕

pie [paɪ] n. 馅饼

Activity 1

Task 1 Scones.

Ingredients:

225 g/8 oz self-raising flour

A pinch of salt

55 g/2 oz butter

25 g/1 oz caster sugar

150 ml/5 fl oz milk

1 free-range egg, beaten, to glaze (alternatively use a little milk)

Directions:

Heat the oven to 220℃/425℉/Gas 7. Lightly grease a baking sheet.

Mix together the flour and salt, and rub in the butter.

Stir in the sugar and then the milk to get a soft dough.

Turn on to a floured work surface and knead very lightly. Pat out to a round 2 cm/¾ in thick. Use a 5 cm/2 in cutter to stamp out rounds and place on a baking sheet. Lightly knead together the rest of the dough and stamp out more scones to use it all up.

Brush the tops of the scones with the beaten egg. Bake for 12 — 15 minutes until well risen and golden.

Cool on a wire rack and serve with butter and good jam and maybe some clotted cream.

Task 2 Peanut Butter Cookies.

Ingredients:

8 tablespoons of plain flour

2 tablespoons of caster sugar

2 tablespoons of crunchy peanut butter

1 free-range egg yolk

50 g/1¾ oz butter, softened

Icing sugar, for dusting

Directions:

Preheat the oven to 180℃/350℉/Gas 4.

Place the flour, sugar, peanut butter, egg yolk and butter into a large bowl and mix together until combined to a smooth dough.

With lightly floured hands, break off evenly sized pieces of the dough and roll into walnut sized balls.

Place the dough balls onto a baking sheet lined with silicon paper or baking parchment and gently press each ball with the back of a fork to flatten slightly.

Place in the oven and bake for 10 minutes, or until just turning golden-brown.

Transfer the cookies to a wire rack and allow to cool for 10 minutes.

To serve, place on a serving plate and dust with icing sugar.

Task 3 Pork Chops with Creamy Bacon Cabbage.

Ingredients:

For the smoky potatoes:

2 baking potatoes, cut into 2.5cm/1 in cubes

4 tablespoons of olive oil

100 g/3½ oz unsalted butter

1 tablespoon of smoked paprika

For the pork chops:

4×200 g/7 oz pork chops

2 tablespoons of olive oil

Salt and freshly ground black pepper

For the creamy bacon cabbage:

300 g/10½ oz smoked bacon lardons

1 head Savoy cabbage, shredded and blanched

125 ml/4 fl oz double cream

Directions:

Bring a large saucepan of salted water to the boil and add the potatoes. Parboil for 4—5 minutes until just tender. Drain and set aside.

Meanwhile, snip vertical cuts in the pork chop fat at 2cm/1in intervals to stop the chops curling up when cooking. Rub the chops with the oil and season well with salt and freshly ground black pepper.

Heat a griddle pan until hot and griddle the pork chops for about 6 minutes on each side, or until cooked through.

Meanwhile, fry the bacon lardons in a frying pan until crisp. Add the shredded and blanched cabbage to heat through before adding the cream. Mix well to coat the cabbage and bacon with the cream.

For the smoky potatoes, heat the oil in a frying pan and fry the parboiled potatoes until crisp. Add the butter and paprika and season well with salt and freshly ground black pepper.

To serve, spoon the creamy cabbage onto four serving plates, top with a pork chop and serve the smoky potatoes alongside.

Activity 2

Task Reading.

Baking is a cooking technique using prolonged dry heat acting by convection, rather than by thermal radiation, normally in an oven, but also in hot ashes, or on hot stones. The most common baked item is bread but many other types of foods are baked. Heat is gradually transferred "from the surface of cakes, cookies and breads to their centre. As heat travels through it transforms batters and doughs into baked goods with a firm dry crust and a softer centre". Baking can be combined with grilling to produce a hybrid barbecue variant, by using both methods simultaneously or one before the other, cooking twice. Baking is related to barbecuing because the concept of the masonry oven is similar to that of a smoke pit.

Baking has been traditionally done at home by women for domestic consumption, by men in bakeries and restaurants for local consumption and when production was industrialized, by machines in large factories. The art and skill of baking remains a fundamental one and important for nutrition, as baked goods, especially breads, are a common food, economically and culturally important. A person whose career is baking is called a baker.

Question 1: What is baking?

Question 2: What is a baker?

Chapter 6

Western Etiquette

Unit 1 Tableware

Learning goals

To know essential information of Western tableware.

To get to know how to set a table.

Vocabulary

etiquette [ˈetɪˌket] n. 礼仪;礼节;规矩

tableware [ˈteɪbəlˌweə(r)] n. 餐具

glassware [ˈglaːsweə(r)] n. 玻璃制品

dishware [ˈdɪʃˌweə] n.(上菜的)餐具

serve [sɜːv] v. 服务;侍候

table setting 工作台紧固

course [kɔːs] n. 课程;讲座;过程;路线;一道(菜)

dinner [ˈdɪnə(r)] n. 晚餐;晚宴;主餐

dining [ˈdaɪnɪŋ] n. 就餐

Dialogue

A: Do you often go to KFC?

B: I've been there only once.

A: I thought all of you youngsters love Western fast food.

B: I'm an exception. I prefer Chinese cuisines.

A: You have some kind of bias on fast food.

B: Though fast food tastes good, it has high calories and little nutrition.

A: Actually Chinese fast food has few fried cuisines, and Western fast food does have some low-calorie and nutritious dishes, such as vegetable soup and salad.

Activity 1

Task Discuss with your partner, and try to describe table settings of different courses.

1. Formal place setting

2. Appetizer course setting

3. Soup course setting

4. Fish course setting

5. Entrée course setting

6. Palate cleanser

7. Main (Releve) course setting

8. Salad course setting

9. Fingerbowl before dessert setting

10. Dessert place setting

11. Dessert course setting

12. Coffee/tea course setting

13. Tea service setting

Activity 2

Task 1 Reading.

Tableware

Tableware is the dishes or dishware used for setting a table, serving food and for dining. Tableware can be meant to include cutlery and glassware. The nature, variety, and number of objects vary from culture to culture, from religion to religion and from cuisine to cuisine.

In the United States, tableware is most commonly referred to as dinnerware. Dinnerware can be meant to include glassware, but not flatware. In Britain, the term crockery is sometimes used for ceramic dishes. In the USA, ceramic dinnerware is often referred to as china. Sets of dishes are often referred to as a table service or service set. Table settings or place settings are the dishes, flatware (cutlery), and glassware used by an individual for formal and informal dining. In the United Kingdom, silver service or butler service are names of methods for serving a meal.

Dinnerware

Dishes are usually made of ceramic materials such as earthenware, stoneware, bone china or porcelain, whatever can be made of other materials such as wood, pewter, silver, gold, glass, acrylic and plastic. Dishes are purchased either by the piece or by set which include either four, eight, or twelve place settings. Individual pieces, such as those needed as replacement pieces for broken dishes, can be purchased from "open stock" inventory at shops, or from antique dealers if the pattern is no longer in production.

Question 1: What is tableware?

Question 2: Now many kinds of dishes are contained?

Table Setting

Table setting or place setting refers to the way to set a table with tableware—such as eating utensils and dishes for serving and eating. The arrangement for a single diner is called a place setting. The practice of dictating the precise arrangement of tableware has varied across cultures and historical periods.

A table setting may have many elements, especially on formal occasions; the long utensil is a lobster pick.

Place Setting

Informal settings generally have fewer utensils and dishes but use a layout based on more formal settings. Utensils are arranged in the order and according to the manner in which the diner will use them. In the West, forks, bread plate, butter knife, and napkin generally are placed to the left of the dinner plate, and knives, spoons, stemware and tumblers, cups, and saucers to the right. (By contrast, formal settings in Greece, Armenia, and Turkey place the fork to the right of the dinner plate.) Sauceboats and serving dishes, when used, either are placed on the table or, more formally, may be kept on a side table.

Table Setting

A formal table setting for one person.

Utensils are placed about an inch from the edge of the table, with all placed either upon the same invisible baseline or upon the same invisible median line. Utensils in the outermost position are used first (for example, a soup spoon and a salad fork, then the dinner fork and the dinner knife). The blades of the knives are turned toward the plate. Glasses are placed an inch or so above the knives, also in the order of use: white wine, red wine, dessert wine, and water tumbler.

Formal Dinner

The most formal dinner is served from the kitchen. When the meal is served, in addition to the central plate (a service plate or dinner plate at supper; at luncheon, a service plate or luncheon plate) at each place there are a bread roll (generally on a bread plate, sometimes in the napkin), napkin, and flatware (knives and spoons to the right of the central plate, and forks to the left). Coffee is served in butler service style in demitasses, and a spoon placed on the saucer to the right of each handle. Serving dishes and utensils are not placed on the table for a formal dinner. The only exception in the West to these general rules is the protocol followed at the Spanish royal court, which was also adopted by the Austrian Habsburg court, in which all flatware was placed to the right of the central plate for each diner.

At a less formal dinner, not served from the kitchen, the dessert fork and spoon can be set above the plate, fork pointing right, spoon pointing left.

In Europe, if many courses are to be served, the table is only laid for soup, fish, and meat. The pudding spoon and fork and the savoury knife and fork are then placed on the table when and as required.

Table Setting

At an informal setting, fewer utensils are used and serving dishes are placed on the table. Sometimes the cup and saucer are placed on the right side of the spoon, about four inches from the edge of the table. Often, in less formal settings, the napkin and/or cutlery may be held together in a single bundle by a napkin ring. However, such objects as napkin rings are very rare in the United Kingdom, Spain, Mexico, and Italy.

Question 1: What is table setting?

Question 2: What different between formal and informal table settings?

Task 2 Further reading.

Chinese Tableware

A Chinese table setting for a group meal

A place setting for a Chinese meal

Chinese table settings are traditional in style. Table setting practices in Japan and other parts of East Asia have been influenced by Chinese table setting customs. The emphasis in Chinese table settings is on displaying each individual food in a pleasing way, usually in separate bowls or dishes. Formal table settings are based upon the arrangements used in a family setting, although they can become extremely elaborate with many dishes. Serving bowls and dishes are brought to the table, where guests can choose their own portions. Formal Chinese restaurants often use a large turning wheel in the centre of the table to rotate food for easier service.

In a family setting, a meal may include a *fan* food, meaning the main dish, and several accompanying side dishes, called *cai* food. The *fan* food is typically a grain, such as rice or noodles. If the meal is a light meal, it will include the staple food and perhaps one side dish. The staple food is often served directly to the guest in a bowl, whereas side dishes are chosen by the guest from serving dishes on the table.

Place Setting

An "elaborate" formal meal would include the following place setting:

Centre plate, about 6 inches in diameter

Rice bowl, placed to the right of the centre plate

A small cup of tea, placed above the plate or the rice bowl

Chopsticks to the right of the centre plate, on a chopstick rest

A long-handled spoon on a spoon rest, placed to the left of the chopsticks

Small condiment dishes, placed above the centre plate

A soup bowl, placed to the left above the centre plate

A soup spoon, inside the soup bowl

A Meal

Meal

Admission, call the owner, that is to start eating.

Take food, not sung too much. After eating food, if not, can be taken. If the greeter at the dish, to be added, to be sent when the receptionist again. If I can not eat or not eat the food, or serve as the master greeter Jia Cai, do not refuse, on the desirability of a small amount of disk, and said "Thank you, enough." Taste of the dishes are not, do not show An embarrassed expression.

To eat refined. Chewing shut up, do not drink soup, not to eat sounds. Such as soup, hot dishes, cold wait before eating, not to blow his mouth. Inside the mouth of the fish bone, the bones do not directly outside the spit, Yanzui napkins, hand (can be used chopsticks to eat Chinese food) out, or lightly fork in the spit, vegetables on tray.

The leftovers of food, utensils used toothpicks should be placed on disk, not to put their table.

Mouth with food, not to speak. Tiya, napkin or hand over the mouth.

Talk

Both for the owner, or Peike guests, and the table should talk to people, especially around next to him. Not only with a few acquaintances or two of the same words. If the neighbor did not know, introduce myself first.

Toast

As the guest of honor at the banquet to participate in foreign, should be used to understand each other's toast, that is why people toast, when the toast, and so on, in order to make the necessary preparations. Clink, and the owner of the guest of honor first touch, many people may indicate a toast at the same time, not necessarily clink. Toast careful not to cross when the clink. And the master chief guest at the speech, toast, the meal should be suspended, to stop talking and pay attention to listening, and do not like to take this opportunity to smoke. Stood when the national anthem is played. The master and guest of honor, then finished with VIP guests clink staff, often to Teachers sprinkle the other table, should an emergency occur, rose to toast. Clink, to pay tribute to each other by sight.

Dinner toast each other, said the friendly, lively atmosphere, but bear in mind that excessive drinking. Drinks too easily slip of the tongue, and even loss, it is necessary to control himself in less than one-third of those who have.

Undress

In social occasions, no matter how hot the weather can not solve the buttons off his clothes in public. Small informal dinner, and invited guests, such as the masters of undress, gentlemen can take off his jacket in the back.

Tea

(Or coffee) tea, coffee, plus get milk, sugar, cup self to join with small mixing teaspoon, teaspoon back into the still small dishes, usually milk, sugar are in full bloom with separate utensils. Drink at the right hand is holding the Cup, the left hand side small dishes.

Fruit

Chili, Apple, do not bite with a whole should be cut into four with a fruit knife, 6, and then peeled with a knife, nuclear, and then eat with their hands, peeling knife-edge when North Korea, cut inside from the outside . First banana peel, cut into small pieces to eat with a knife. With a knife and cut into pieces to eat oranges, orange, lychee, longan and so on can eat Bole Pi. The rest, such as watermelon, pineapple, etc. , usually go into skin yuan, can be used when eating a fruit knife and cut into small pieces with fork food.

Shui Yu

In the banquet, the chicken, lobster, fruit, in some cases sent a small Shui Yu (Tongpen , crystal bowls or Boli Gang), a floating rose petal water or lemon slices for the use of hand-washing (some people have mistaken for beverages, As a result become a joke). Wash hands when they take turns damp fingers gently Shuanxi, and then use a small napkin or towel dry.

Memorial items

Some of the master for each person attending or have a small souvenir of flowers. At the end of the banquet, call the owner to bring guests. In such, it can be said that it commended the 12 small gifts, but do not have to solemnly said. In some cases, foreign visitors, the dinner menu is often taken as a souvenir, and sometimes I please with those who signed the menu as a souvenir. In addition to the special master to indicate things as souvenirs, a variety of entertainment products, including candy, fruit, cigarettes and so on, are not taken away.

Thanks

Sometimes in the private sector attended the banquet activities, often my business card or memo said.

Will buffet, cocktail buffet vegetables take, cocktail reception, greeter serve, not to get a gun, I have to be sent to the front of the airport project. Did not get around the first time, not to rush to get their second. Do not dish around the table next to the check End Tuikai that, in order to let other people get.

The use of tableware

Chinese food is the major bowls, chopsticks, is the Western knives, forks, plates. Dinner is usually a foreigner to eat Chinese food, Chinese food is also for the West to eat more, before we go, knife and fork set. The knife and fork is the use of his right hand with a knife, hold the left hand fork, cut into small pieces of food, and then sent a cross inside the mouth. Europeans do not use changing hands, from cutting food were sent to hold the left hand fork. Americans, after cutting, put down the knife, fork right hand holding the entrance to send food. When dinner knife and fork in accordance with the order from outside access inside. After each course, emissions will be set within feet close together and knife and fork to eat that. If finished, then put into a character or a cross placed, knife-edge to inside. Chicken, lobster, indicated by the master, can be torn by hand to eat, or can be cut meat knife and fork, cut into small pieces to eat. Cut with bone or hard shell of meat, meat fork must fork in prison, knife-edge close to cross under the cut, so as not to slide open. Vegetable, be careful not to hit too much force plate and sound. Is not easy to cross the food, or food is not easy on the fork, gently push it available Dao fork. In addition to the soup, do not have to spoon feeding. With deep soup plate or small bowl full bloom, when to drink with a spoon scoop from the inside out into the mouth, is about to do drink, can be set to hold out a little. Eat with the smell of food, such as fish, shrimp, game, etc. equipped with a lemon, juice will hand out food in the drip, to smell out.

Unit 2 Table Manners

Learning goals

To know essential information of table manners.

Vocabulary

formal [ˈfɔːml] *adj*. 正式的;正规的

table manner 餐桌规矩;进餐礼节

napkin [ˈnæpkɪn] *n*. 餐巾

main course 主菜

host [həʊst] *n*. 主人

hostess [ˈhəʊstəs] *n*. 女主人

tooth pick 牙签

tongue [tʌŋ] *n*. 舌头

Dialogue

(*H*=*Hostess*, *G*=*Guest*.)

H: Would you like to have some more chicken?

G: No, thank you. The chicken is very delicious, but I'm just too full.

H: But I hope you save some room for the dessert. Mary makes very good pumpkin pies.

G: That sounds very tempting. But I hope we can wait a little while, if you don't mind.

H: Of course. How about some coffee or tea now?

G: Tea, please. Thanks.

Activity 1

Task Translate the following Chinese into English according to the picture.

原则上男主宾(gentleman of honor)坐在女主人(hostess)右边,女主宾(lady of honor)坐在男主人(host)右边,而且多半是男女相间而坐,夫妇不在一起,以免各自聊家常话而忽略与其他宾客间的交际。

Activity 2

Task 1 Reading.

Table Manners

All the food a person intends to eat is put on a large plate in front of him at the beginning of the meal (分食制). If one finishes the food on his plate and wants some more, he will get a second helping.

Some families will say grace before eating. It is common with Christian families but not all families.

Soups should be drunk with spoons. When eating soup, don't make noise and don't lift you bowl.

Toasting at meals is uncommon so you can drink when you feel thirsty.

The waiter will serve the dishes to the guests one by one. When he stands on your left, it is your turn to take the food. Otherwise, it is not your turn.

As to the seating, there are places of honor. And the honored guests sit on the right side of the host and hostess. And the host and hostess often sit at the two ends of the ordinary rectangular(长方形的) tables. People of the same sex avoid sitting together.

If a dessert is to be served, the table will be cleared first. It is polite for the guests to let the hostess sit down before you take the first bite of the dessert.

If there is liquor or wine on the table, drink it after all the glasses of the guests are filled.

Hold the knife with your right hand and the fork, left. After cutting the food, put your knife on the table and eat the food with your fork in your right hand.

Slurping (啧啧) and burping (打嗝) are considered as extremely rude. If you have to, say "excuse me".

It is also considered rude to sing at the table, to put your elbows on the table or to chew your food with you mouth open, to talk with your mouth full, or to make loud noise by striking the utensils against the plate.

Avoid picking your teeth and if you have to, cover your mouth with the napkin.

At the dinner table, it is bad manners to remain silent all the time. But don't talk when you are chewing. Don't swing your knife and fork while speaking.

Question 1: What do Christian families do before eating?

Question 2: Which hand should you hold the knife?

Task 2 Further reading.

African Cuisines

Cuisine of Egypt

The Egyptian cuisine consists of local culinary traditions such as Ful Medames, Kushari and Molokhia, while sharing similarities with food found throughout the eastern Mediterranean like kebab and falafel. Most Egyptians, perhaps, consider Ful Medames, or mashed fava beans, to be their national dish. Ful is also used in making Ta'miyya or Falafel. Bread accompanies most Egyptian meals; local bread is called Eish Masri or Eish Baladi. Ancient Egyptians are known to have used a lot of garlic and onion in their everyday dishes. Fresh mashed garlic with other herbs is used in spicy tomato salad and is also stuffed in boiled or baked aubergines (eggplant). Garlic fried with coriander is added to Mulukhiyya, a popular green soup made from finely chopped leaves. Fried onions are added to Kushari, a dish consisting of brown lentils, macaroni, rice, chickpeas and a spicy tomato sauce.

Cuisine of South Africa

The cuisine of South Africa reflects the diversity of the various ethnic groups that make up the population of South Africa. The variety of different culinary traditions combined with the great variety of fresh food present a cuisine that caters for every taste sensation. Although the historical traditions are evident in the cuisine, modern food trends also influence its development.

South African cuisine is strongly influenced by Dutch, German, English, French, Malaysian, Portuguese, Indonesian, Indian, Native African and even Asian cuisines. As a result, we have a great variety of dishes.

Asian Cuisines

Cuisine of China

The Chinese cuisine is actually not a single entity, but is instead made up of the individual cuisines of many provinces and ethnics groups of China. So combined together, there are perhaps thousands of types all over the China. That is why Chinese have a famous old saying "Cuisine is heavenly in first priority" (民以食为天). However, there are some common ingredients and philosophies about most Chinese foods.

First and foremost, the Chinese believe that having a knife at the table is barbaric. This means dishes are usually prepared in portions to be eaten directly with chopsticks or spoons. Second, Chinese foods are usually served warm and well-cooked.

Cuisine of Vietnam

The Vietnamese cuisine (ăm thăc in Vietnamese: ăm: drink and thăc: food) is known for its common use of fish sauce, soy sauce and hoisin sauce. Vietnamese recipes use many vegetables, herbs and spices, including lemon grass, lime, and kaffir lime leaves. Throughout all regions the emphasis is always on serving fresh vegetables and/or fresh herbs as side dishes along with dipping sauce. The Vietnamese also have a number of Buddhist vegetarian dishes. The most common meats used in the Vietnamese cuisine are pork, beef, prawns, various kinds of tropical fish, and chicken. Duck and goat/lamb are used much less.

European Cuisines

Cuisine of the United Kingdom

The British cookery is sometimes regarded as a figure of fun but actually it ranks with the best cuisines in the world. The British cooking, however, is not just traditional "roast beef and Yorkshire pudding". It reflects and incorporates something from all the cultures and countries Britain has had contact with throughout her long history. Indeed the British cooking has inspired many culinary traditions around the world.

Cuisine of France

The French cuisine is characterised by its extreme diversity. Despite France's history of political and cultural centralization around its capital Paris, each region has its own distinctive specialities: Cuisine from North-West France uses butter and cream; the Provençal cuisine (from the southeast) favours olive oil and herbs; and the eastern French recipes are reminiscent of the German cuisine, including sausages, beer and sauerkraut. Wine and cheese are an integral part of the French cuisine, both as ingredients and accompaniments.

Mediterranean Cuisines

Cuisine of Italy

The Italian cuisine has a tradition of dishes based on wheat products (such as bread and pasta), vegetables, cheese, fish, and meat, usually prepared in such a manner as to preserve their ingredients' natural qualities, appearance, and taste.

This kind of cuisine puts a stress on lightness and healthy nutrition with natural not processed foods, and tends to vary greatly not only with the seasons but also between the various regions of the country: Mountainous regions have dishes rich in proteins, and prefer meat, butter, and cheese, while seaside regions have dishes rich in vegetables and fish.

In this way, the cuisine is born of the people, the territory and the seasons, and is not pulled out of nowhere for no rhyme or reason.

Cuisine of Spain

The cuisine of Spain is one of the most intriguing the world has to offer. The diverse landscape, climate, and culture of Spain, combined with its rich history, create a cuisine unlike any other. The turbulent history of the Iberian Peninsula had a great influence on the cuisine of Spain, as the Romans, the Visgoths, and the Moors all left their culinary marks. The post-Franco Renaissance that encouraged richness and diversity in all aspects of life is demonstrated through the unique Spanish way of cooking. Spain is made up of 17 autonomous regions, each differing in natural resources, creating a variety of specialty dishes specific to each region. In general, the Spanish cuisine relies on the Mediterranean flavors—garlic, olive oil, and fresh herbs. The Spanish cuisine is sophisticated, yet basic, unadorned and elegant.

American Cuisines

Cuisine of Trinidad and Tobago

The cuisine of Trinidad and Tobago is indicative of the blends of Amerindian, European, African, Indian, Chinese, Creole, and Lebanese gastronomic influences, and is also more similar to the cuisine of Guyana than most other Carribbean countries.

A representative main dish from Trinidad and Tobago is callaloo, a creamy and spicy sauce made of dasheen leaves, okra, coconut milk, pumpkin and chandon benit (French thistle or fitweed). Crab and callaloo is generally considered a national dish of Trinidad and Tobago; it is often prepared for Sunday lunch. Pelau, a rice—based dish, is also a standard dish in Trinidad and Tobago. Another popular dish is roti—this is of East Indian origin and consists of curried potatoes, channa (chickpeas) and meat wrapped in dalphurie (similar to pita bread). Other local dishes include coo coo, sancoche, macaroni pie and oil down.

美国的各级厨师执照

1. Certified Culinarian or Pastry Chef　合格的厨师或糕点师傅

厨师必须具有食物烹调、烘焙、营养、安全和公共卫生基本知识。

2. Personal Certified Chef　合格的私人厨师

私人厨师像专业厨师一样,需要数年的工作经验,负责菜单规划、财务管理和私人业务的经营事项。

3. Certified Sous Chef　合格的厨师助理

厨师的助理像高级厨师或料理厨师,最少需要两年的食物烹饪管理经验,对食物安全和公共卫生、烹饪营养和监督管理方面,亦具有充分的知识。

4. Certified Working Pastry Chef　合格、有经验的糕点师傅

有经验的糕点师傅,管理糕点部分或轮值食品服务的管理,负责烘焙蛋糕、派、饼干、面包、卷饼等等的烘焙食品。

5. Certified Chef de Cuisine　合格的料理厨师

料理厨师负责监督食物的烹调,需有食品服务经营至少三年的管理经验。有时候负责作决策,对食品安全、公共卫生、烹饪营养和监督管理,需要有良好的知识。

6. Certified Secondary Culinary Educator　合格的烹饪助教

合格的烹饪助教需对烹饪技术、教育发展、食品安全、公共卫生和烹饪营养皆具有良好的知识,并且在高职或职业烹饪训练中心担任过教师。

7. Certified Culinary Educator　合格的烹饪老师

合格的烹饪老师任教过高职、专科学校或军中的烹饪训练单位,具有充分的烹饪知识,能教授管理食品服务的开发、实践、经营和评估等课程。

8. Personal Certified Executive Chef　合格的高级私人厨师

具有高级的烹饪技巧,烹饪的工作经验至少六年以上,能对不同的客户提供烹饪服务,并负责菜单的设计、行销、财务管理和经营决策等。

9. Certified Executive Chef/Certified Executive Pastry Chef　合格的高级厨师/高级糕点师

高级厨师/高级糕点师是部门的首长，负责领导餐厅或饭店之厨房部门的全部事宜，监督经理，负责各个部门的协调、菜单设计、预算控制和存货记录。

10. Certified Master Chef/Certified Master Pastry Chef 合格的主厨/首席糕点师

主厨/首席糕点师需具有最高级的专业烹调知识和技能，负责教导、监督、领导食品部门的全部的安全人员。

中国台湾的厨师执照

1. Certified C-level Chef　丙级厨师

A C-level chef is an entry level chef who has completed a basic traning session of food preparation, safety, and sanitation.

He/she has to be above the age of 15 and has a diploma of junior high school education.

A certified C-level chef is qualified to work at the food stand, cafeteria, catering, and school/factory canteen.

丙级厨师是指已完成食物烹调、安全和公共卫生之基础训练的课程，必须年满 15 岁，可以成为小吃摊子、自助餐厅、酒席、学校餐厅/工厂餐厅的合格厨师。

2. Certified B-level Chef　乙级厨师

A B-level chef is the higher degree chef in Taiwan. He/she should have in total 1,600 hours training in the relevant subjects or at least two years of working experience as a chef after obtaining the C-level chef certificate. This person usually works as an executive chef, a culinary educator or a restaurant owner.

乙级厨师需具备较高的学位，必须修相关科目至少 1600 个小时，或是取得丙级厨师执照后有两年以上的厨师工作经验。可担任高级厨师、烹饪老师或经营自助餐厅。

Appendix

Key to Exercises

Chapter 1

Unit 1

Activity 1

Task 1

1. executive chef
2. roast cook
3. larder chef
4. vegetable cook
5. grill cook
6. soup cook
7. pastry chef
8. relief cook

Task 2

1. relief cook
2. aboyeur
3. breakfast cook
4. potman
5. roast cook
6. pantryman
7. butcher

Activity 2

Task 1

1. What do you do in the kitchen, please?
2. How long does a commis/apprentice work every day in the kitchen?
3. The vegetable chef is mainly responsible for cooking vegetable dishes.
4. How did you find the potman's job?
5. Brian is a butcher. He has to butcher different kinds of poultries in the kitchen.

Unit 2

Activity 1

Task

1. put out
2. wipe up
3. wash
4. sanitize
5. clean

Activity 3

Task 1

1. 冰箱;冰柜
2. 冷菜房
3. 屠宰房
4. 面点房
5. 饮料冷库
6. 厨师长办公室
7. 热菜房
8. 厨具清洗房

9. 蔬菜配菜间
10. 鱼类菜品制作房
11. 食器清洗房
12. 厨房储藏间
13. 备菜间

Unit 3

Activity 1

Task 1

1. divide the dough
2. roll the dough
3. flatten the dough
4. knead the dough
5. shape the dough
6. fold the dough
7. puff pastry
8. line the mold

Activity 2

Task

1. Please flatten the pastry with a rolling pin.
2. The little boy kneaded the dough into a ball.
3. Garnish the fish with cucumber slices.
4. Please stretch the dough and then divide it into three parts.
5. I want to frost the cake with icing.
6. Could you give me some flour? I want to make some puff pastries.

Activity 3

Task

English	French	Description
sauté chef	saucier	Responsible for all sautéed items and their sauce. This is usually the highest stratified position of all the stations.

续 表

English	French	Description
fish chef	poissonnier	Prepares fish dishes and often does all fish butchering as well as appropriate sauces. This station may be combined with the *saucier* position.
roast chef	rôtisseur	Prepares roasted and braised meats and their appropriate sauce.
grill chef	grillardin	Prepares all grilled foods; this position may be combined with the *rotisseur*.
fry chef	friturier	Prepares all fried items; this position may be combined with the *rotisseur* position.
vegetable chef	entremetier	Prepares hot appetizers and often prepares the soups, vegetables, pastas and starches. In a full brigade system a *potager* would prepare soups and a *legumier* would prepare vegetables.
pantry chef	garde manger	Responsible for preparing cold foods, including salads, cold appetizers, *pâtés* and other *charcuterie* items.
butcher	boucher	Butchers meats, poultry and sometimes fish. May also be responsible for breading meats and fish.
pastry chef	pâtissier	Is qualified in making baked goods such as pastries, cakes, biscuits, macarons, chocolates, breads and desserts. Pastry chefs are specialized in cakes, patisseries or bakeries by making wedding, cupcakes, birthday and special occasion cakes. In larger establishments, a pastry chef often supervises a separate team in their own kitchen or separate shop.

Unit 4

Activity 1

Task 1

1. scale the fish
2. bone the fish
3. flatten the meat
4. cut the beef

5. cut the fish
6. slice the fish
7. rinse the fish
8. tie up the meat

Activity 2

Task 1

1. Scale the fish before cooking them.
2. Please bone the beef and flatten the meat.
3. Cut open the stomach of the fish. Then take out the guts.

Task 2

3 4 2 5 1

Unit 5

Activity 1

Task 1

1. beat the eggs
2. mash the potatoes
3. clean/wash the vegetable
4. cube the carrots
5. mince the meat
6. shred the cucumber
7. remove the seeds
8. slice the tomatoes
9. sprinkle the cheese
10. refrigerate the food
11. spread the jam
12. sharpen the knife
13. sift the flour
14. weigh the orange

Activity 2

Task 1

1. Give that floor a good hard scrub.
2. Mince the lean pork and marinate it with seasoning.
3. Please remove the seeds of the papaya.
4. Mary asked her husband to preheat the oven to 180 degrees Celsius.
5. He sprinkled vinegar on his fish and chips.

6. The waiter put a squeeze of lemon in my drink.
7. It won't grind down any finer than this.

Task 2

1. g 2. h 3. f 4. j 5. i 6. a 7. c 8. e 9. d 10. b

Chapter 2

Unit 1

Activity 1

Task 1

1. colander
2. skimmer
3. sieve
4. chinois
5. frying basket
6. conical strainer

Activity 2

Task 1

1. The top layer of the oil in the deep fryer now has a lot of small pieces of floating food.
2. Should I sift the flour with a sieve?
3. What do you use the conical strainer for?
4. I want to clean the surface of oil with a skimmer.
5. I need a chinois. Could you hand one to me?

Task 2

1. colander

Descriptions:

A colander is a bowl-shaped kitchen utensil with holes in it to be used for draining food such as pasta or rice.

2. frying basket

Descriptions:
A frying basket is used to strain oil of frying food.

Unit 2

Activity 1

Task 1

1. roasting fork
2. skewer
3. knife set
4. chopper
5. ceramic knife
6. boning knife
7. fish scissor
8. fruit knife
9. oyster knife
10. chopping board
11. cheese knife
12. poultry shear
13. peeler

Activity 2

Task 1

1. What is this funny little knife for, please?
2. Can I use the knife set to open the lid of this jar of chutney?
3. A boning knife is used to remove bones from meat.
4. Could you peel an apple for me?
5. How many different kinds of knives are included in a knife set?
6. What is the difference between ordinary scissors and poultry shears?

Task 2

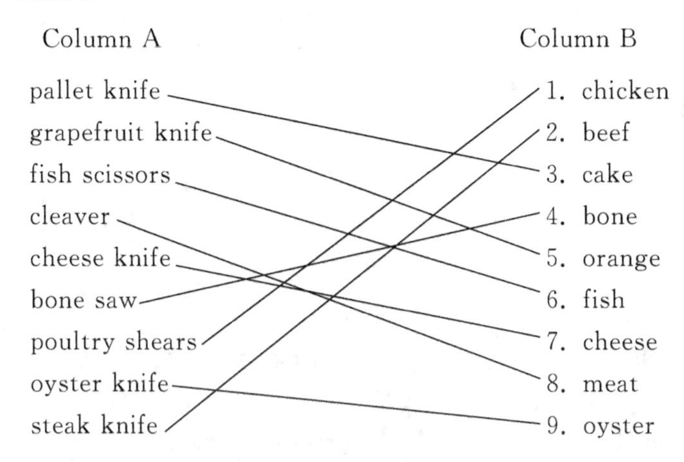

Unit 3

Activity 1

Task 1

1. pallet knife
2. spatula
3. slotted spoon
4. basting brush
5. mixing bowl
6. mallet
7. ladle
8. soup tureen

Activity 2

Task 1

1. That's a nice birthday cake. What should we do next?
2. Please put a layer of icing sugar on the cake with a palette knife.
3. —What do you call this spoon?

   —This is called a slotted spoon.
4. Let's put it into the soup tureen.
5. Use the mallet to flatten the meat into escolopes.
6. Can I mix the mayonnaise and the tuna in this mixing bowl?

Task 2

1. e 2. b 3. f 4. g 5. c 6. a 7. d

Unit 4

Activity 1

Task 1

1. sauce pan
2. bain marie
3. stew pan
4. sauté pan
5. frying pan
6. oven
7. roasting tray
8. pressure cooker

Activity 2

Task 1

1. Could you flatten the meat with a cutlet bat?
2. A bain marie is used to keep the dishes warm.
3. Watch out! This pressure cooker is very hot.
4. What should I do with this roasting tray?
5. Please stew the chicken in a saucepan.
6. Never fill a pot to the top. And when you lift the lid, do it very slowly and carefully.

Task 2

1. frying pan

Descriptions:

A frying pan is used to fry meat, fish, eggs, etc.

2. pressure cooker

Descriptions:

A presssure cooker is used to cook food quickly by steam under high pressure.

3. oven

Descriptions:

An oven is used to roast large pieces of meat, pizza, potatoes, etc. by covering the surface of the food with oil.

Unit 5

Activity 1

Task 1

1. whisk/egg beater
2. steel
3. wooden chopping board
4. plastic chopping board
5. hot pad
6. hot glove
7. mallet
8. blender
9. tenderizer
10. egg boiler
11. toaster
12. dish washer
13. microwave
14. steamer
15. casserole
16. grinder
17. cake cutters
18. tart tin
19. grater
20. electric cook tops
21. outdoor barbecue
22. gas burner
23. kettle
24. coffee machine
25. refrigerator
26. food processor
27. rice cooker
28. electric mixer
29. trash compactor
30. dryer
31. range cooker
32. ventilation

Activity 2

Task 1

1. Please put the oranges into the blender.
2. What do you use an electric slicer for?
3. Cover it with a cloche.
4. I'll clean it with a sponge and some soapy water.

5. Will the bowls be cleaned by using a dish washer?
6. Use a hot glove to take a pan out of the oven.

Task 2

Cakes: cake cutters/tart tin
Toasts: toaster
Egg wash: whisk/egg beater
Chicken soup: casserole
Steamed bun: steamer
Meat paste: tenderizer
Apple juice: blender

Chapter 4

Unit 1

Activity 1

Task

1. asparagus
2. beetroot
3. burdock
4. cabbage
5. carrot
6. cauliflower
7. celery
8. corn
9. cucumber
10. garlic
11. green pepper
12. lettuce
13. mushroom
14. okra
15. onion
16. parsley
17. pea
18. pumpkin
19. eggplant
20. spring onion
21. tomato
22. jew's-ear
23. mustard blue
24. red pepper
25. spinach
26. water chestnut

27. taro
28. towel gourd
29. lotus root
30. potato

Activity 2

Task 2

1. This salad is made of apples, pears, potatoes and celery.
2. Divide the cauliflower into florets and wash them thoroughly.
3. Could I have cauliflower instead of mashed potatoes?
4. Boil 3 cups of water,add salt and sugar,and braise asparagus.
5. Can you taste the garlic in this stew?
6. The eggplant is a kind of purple vegetable.
7. They crush the olives with a heavy wooden press.

Unit 2

Activity 1

Task

1. blackberry
2. cherry
3. fig
4. grape
5. grapefruit
6. lemon
7. melon
8. peach
9. kiwi
10. plum
11. apple
12. coconut
13. loquat
14. star fruit
15. pitaya
16. papaya
17. durian
18. persimmon
19. greengage
20. longan
21. mango
22. cantaloupe
23. orange
24. pear
25. pineapple
26. litchi
27. strawberry
28. pomelo

29. pomegranate
30. watermelon
31. apricot
32. tangerine
33. blueberry
34. avocado

Activity 2

Task

1. The berry is nutritious and has a sweetish taste.

2. The rich and creamy avocado has been called the "chocolate" of fruits.

3. The mango ice cream is pretty good.

4. All you have to do is skin and core a pineapple.

5. Slice a kiwifruit and combine it with turkey, papaya, almond slivers and spinach leaves for a cool salad.

6. The fish is garnished with slices of lemon.

7. It seems that a blueberry pie is being cooked in the kitchen.

8. He spread some strawberry jam on his toast.

Unit 3

Activity 1

Task

1. thyme
2. basil
3. ginger
4. sage
5. black pepper
6. nutmeg
7. saffron
8. cinnamon
9. star anise
10. fennel
11. mint
12. rosemary
13. bay leaf
14. liquorice
15. pepper
16. geranium
17. clove
18. dried tangerine
19. angelica dahurica
20. rhubarb

21. sand kernel
22. pepper
23. ketchup
24. maple syrup

Activity 3

Task 1

1. Fish gravy is a traditional fermented condiment in coastal areas.
2. Then, you put some condiment and a little salt into the noodles.
3. Sashimi is usually seasoned with wasabi.
4. This meat should be seasoned with salt and mustard.
5. In France, mustard seeds are soaked and then grinded to a paste.
6. Chutney can be mixed with any Indian dish to create a different taste.
7. I would like some ham, sausage, mushrooms, onions, black olives, and pineapples for the topping.
8. Salt is a common food preservative.
9. Nutmeg is usually used as flavoring in food.
10. The special test of this soup is due to the saffron.

Task 2

For Chinese, nutmeg is widely known as a kind of spice. In Europe, people have many ways to use nutmeg. The British put it in the rice cakes, tarts and milk; Frenchmen usually put nutmeg in the pastry, meat pies, cakes and sausage for seasoning; the Italians especially love to add the nutmeg sauce to veal steak; the Dutch especially love to add nutmeg to the stew.

Unit 4

Activity 1

Task

1. chicken
2. pork
3. minced steak
4. beef

5. mutton
6. duck
7. goose meat
8. snake meat
9. bacon
10. ham
11. sausage
12. goose liver

Activity 3

Task 1

1. I want to eat water celery soup and beef cutlet.
2. Break three eggs, please, and mix them with salt.
3. He cut the veal into cutlets.
4. He put a fillet steak in the bun.
5. Cut a thin cut of beef from the brisket.
6. Transfer the brisket, fat side up, to a large roasting pan.
7. It's a boneless steak cut from the tenderloin of beef.

Task 2

Rare/Very rare: Fried (Grilled) for no more than 3 minutes. The appearance of the steak is like to be barbecued, but the inside is almost cold, almost like without heat. When you cut it, there will be blood oozed. It tastes very tender and juicy.

Rare: Fried (Grilled) for no more than 4 minutes. The appearance of the steak is like to be barbecued, but the inside is original red, tasting a little hot. When you cut it, there will be blood oozed. It tastes very tender and juicy.

Medium rare: Fried (Grilled) for 6—8 minutes. The appearance of the steak is like to be barbecued. It has been heated inside, tasting a little hot. And the meat looks red. When you cut it, there will be blood oozed. It tastes very tender and juicy.

Medium: Fried (Grilled) for 8—10 minutes. The appearance of the steak is dark brown. And the inside looks pink, while the outside looks like to be barbecued. When you cut it, there will be dark brown gravy oozed. You need to chew it.

Medium well: Fried (Grilled) for 10—12 minutes. The appearance of the steak is dark brown. And the inside looks a little red, while the outside looks like to be barbecued. When you cut it, there will be dark brown gravy oozed. It tastes very tender and juicy. You need to chew it.

Well done: Fried (Grilled) for 12—15 minutes. The appearance of the steak has obvious grill marks. The whole steak is heated through. And the inside looks dark brown. You need to chew it.

Unit 5

Activity 1

Task 1

1. salmon
2. sea cucumber
3. lobster
4. sea prawns
5. oyster
6. shell
7. crab
8. cod
9. tuna
10. plaice
11. mussel
12. king prawn
13. grouper
14. sardine

Activity 2

Task 2

1. The cod was salted away for future use.
2. Haddocks are usually baked, but sometimes are roasted with lots of butter.
3. Atlantic cods and haddocks are usually split and boned for cooking.
4. Japan is a major tuna consumer.
5. Here are the Fried Carp and the Roast Beef, sir.
6. The monkfish was very delicate and great with the orange zest.
7. The Japanese like to take raw salmon.
8. The edible flesh of tuna is often canned or processed.

9. Rinse and drain the scallop, cut it in half, and then slice.

Chapter 5

Unit 1

Activity 2

Task 1

1. Scoop the sweet melon out to balls.
2. Flatten the top of the melon balls.
3. Roll up the Parma ham.
4. Place the ham rolls on top of the melon balls.

Task 2

1. Separate egg yolk and egg white, and mince them.
2. Mix the minced egg yolk and sour cream together, and put it in a plastic ring.
3. Then put the minced egg white on the mixture.
4. Put the fine onion on the egg white.
5. Place the caviar over the top.
6. Sprinkle chopped chive over the dish, and the dish is done.

Unit 2

Activity 1

Task

1. crab Louie salad 路易蟹肉沙拉
2. green salad 田园沙拉
3. potato salad with egg and mayo 美乃兹土豆鸡蛋沙拉
4. pasta salad 意粉沙拉
5. tuna salad 金枪鱼沙拉

6. green papaya salad 青木瓜沙拉
7. Caesar salad 恺撒沙拉
8. chef salad 厨师长沙拉
9. lobster soup 龙虾浓汤
10. vegetarian minestrone soup 意大利蔬菜浓汤
11. cream pumpkin soup 奶油南瓜汤
12. French onion soup 法式洋葱汤

Unit 3

Activity 1

Task

1. cheesecake 乳酪蛋糕
2. poutine 肉汁奶酪薯条
3. raclette 拉克莱特干酪
4. mozzarella 炸马苏里拉奶酪条
5. Swiss fondue 瑞士乳酪火锅

Unit 4

Activity 1

Task

1. bouillabaisse 浓味鱼肉汤
2. ceviche 酸橘汁腌鱼
3. crab stick 蟹棒
4. fish and chips 鱼和炸土豆条
5. fishball 鱼丸
6. fish chowder 鱼肉巧达汤
7. gefilte fish 鱼饼冻
8. paella 西班牙海鲜饭

Unit 8

Activity 1

Task

1. flourless chocolate cake 巧克力蛋糕(无面粉)
2. baked custard 烘焙蛋糕

3. strawberry ice cream 草莓冰淇淋
4. cookies 曲奇
5. chocolate cake 巧克力蛋糕
6. pie 派
7. chocolate 巧克力
8. cheesecake 乳酪蛋糕
9. Czech and Slovak appetizer or snack 捷克斯洛伐克点心
10. Japanese zensai 传统日式点心

Chapter 6

Unit 2

Activity 1

Generally speaking, the gentleman of honor should sit on the right of the hostess. The lady of honor should sit on the right of the host, and mostly, a lady and a gentleman should sit beside each other. Couples do not sit together, so that guests can have more communication with others.

参考文献

[1] 陈亚丽. 厨房英语(第三版)[M]. 北京:旅游教育出版社,2006.
[2] 孙诚. 烹饪英语[M]. 北京:高等教育出版社,2009.
[3] 邢怡. 烹饪英语[M]. 上海:上海交通大学出版社,2010.
[4] 赵丽. 烹饪英语[M]. 北京:北京大学出版社,2010.
[5] BAIRD E. The complete Canadian living cook book [M]. Toronto: Random House of Canada Limited, 2001.
[6] LE CORDON BLEU. Complete cook [M]. San Diego: Thunder Bay Press,2002.
[7] CARRIER R. Feasts of Provence [M]. London: Weidenfeld and Nicolson,1992.
[8] COADY C. Real chocolate [M]. London: Whitecap Book,2003.
[9] DAVIS P. English for food, beverages and food production [M]. Taipei: Cosmos Culture Ltd. ,2008.
[10] HEMPHILL I. The spice and herb Bible [M]. Toronto: Pan Macmillan Australia Pty Limited, 2000.
[11] MAJURE R. English for food production [M]. Taipei: Cosmos Culture Ltd. ,2011.
[12] VAILEY R. Baking: easy &elegant [M]. Bielefeld: HPBook, Inc. , 1982.
[13] PAWSON J, BeLL A. Living and eating [M]. New York: Potter, 2001.
[14] STEWART A. The flavours of Canada [M]. Vancouver: Raincoast,2000.
[15] TROTTER C. Charlie Trotter's vegetables [M]. Berkeley: Ten Speed Press,1996.
[16] VERGE R. Entertaining in the French style [M]. New York: Steward, Tabori & Chang Inc. , 1986.

后　记

本书是以编者多年烹饪英语教学研究、课堂实践的校本教材为基础的烹饪专业英语教材。本教材符合高职教育教学特点，紧密结合烹饪行业，融入了行业特性和语言交流的实用功能。通过本教材的教学，能够达到使学生与外籍厨师就专业问题进行交流，并为日常生活交流打下良好基础的目的。由于编者具有英语国家的教育背景，在编写和教学过程中融合了国外语言教育的先进理念，突出课堂教学的互动，注重培养学生的听说能力。本教材适合高职教育烹饪及相关专业使用，也可作为各大涉外宾馆、酒店员工的英语培训教材或自学教材使用。

本书由六章组成，分别是 Kitchen Introduction and Verbs，Kitchenware，Conversations in the Kitchen，Food Materials，Recipes 及 Western Etiquette。每一章分为几个单元，每个单元都由 Learning goals，Vocabulary，Dialogue，Activities 组成。各个单元的内容都围绕设定的主题展开，贴近最真实的教学场景和教学设备，从厨房介绍、餐具、食品、礼仪和模拟厨房对话等方面入手，围绕烹饪工作者的岗位特性及工作中所需要运用的英语知识，注重语言技能应用和职业性质的结合，使高职院校的学生真正做到学以致用。

本书的编写注重实践，图文并茂，内容全面。本教材的目的不仅在于培养学生的英语语言应用能力，也在于培养学生的实践能力。课堂教学中，建议教师以引导的方式教导学生，不仅仅是教师的单纯讲解，更要通过以学生为主体的互动方式，培养学生的独立思考能力和动口能力，融会贯通。

本书由浙江旅游职业学院烹饪系教师华路宏、外语系教师钟文担任主编。第一至三章由钟文编写，第四至六章由华路宏编写，全书由华路宏统稿。编写过程中，编者多次与杭州西溪喜来登酒店、浙旅集团机场大酒店等酒店行政总厨沟通，获取了丰富的第一手资料。其间，浙江旅游职业学院副院长徐云松教授、烹饪系主任戴桂宝老师及其他资深专业教师对本书的出版给予大力支持和指导，浙江工商大学出版社工作人员付出了辛苦劳动，编写过程中专业学生协助整理了资料，在此表示衷心感谢！

由于时间仓促，编者水平有限，书中难免有不当之处，恳请专家和读者不吝赐教。

编　者

2018 年 1 月